Polynomial Based Iteration Methods for Symmetric Linear Systems

Polynomial Based Iteration Methods for Symmetric Linear Systems

Bernd Fischer
Institute of Mathematics, Medical University of Lübeck, Germany

WILEY–TEUBNER SERIES
ADVANCES IN NUMERICAL MATHEMATICS
Editors
Hans Georg Bock   Wolfgang Hackbusch
Mitchell Luskin   Rolf Rannacher

Springer Fachmedien Wiesbaden GmbH

Copyright © 1996 Springer Fachmedien Wiesbaden
Originally published by B.G. Teubner in 1996

National Chichester (01243) 779777 *National* Stuttgart (0711) 789 010
International +44 1243 779777   *International* +49 711 789 010

All rights reserved

No part of this book may be reproduced by any means,
or transmitted, or translated into a machine language
without the written permission of the publisher.

Other Wiley Editorial Offices

Brisbane · Toronto · Singapore

Other Teubner Editorial Offices

Die Deutsche Bibliothek - CIP-Einheitsaufnahme
Fischer, Bernd:
Polynomial based iteration methods for symmetric linear
systems / Bernd Fischer.

(Wiley-Teubner series in advances in numerical mathematics)
   ISBN 978-3-663-11109-2   ISBN 978-3-663-11108-5 (eBook)
   DOI 10.1007/978-3-663-11108-5

WG: 27   DBN 94.726798.0   96.04.03
3909   nh

British Library Cataloguing in Publication Data
A catalogue record for this book is available from the British Library

ISBN 978-3-663-11109-2
Produced from camera-ready copy supplied by the authors

This book is printed on acid-free paper responsibly manufactured from sustainable forestation, for
which at least two trees are planted for each one used for paper production.

To Henning and Martin

Preface

Any book on the solution of nonsingular systems of equations is bound to start with
$$Ax = f,$$
but here, A is assumed to be symmetric. These systems arise frequently in scientific computing, for example, from the discretization by finite differences or by finite elements of partial differential equations. Usually, the resulting coefficient matrix A is large, but sparse. In many cases, the need to store the matrix factors rules out the application of direct solvers, such as Gaussian elimination in which case the only alternative is to use iterative methods.

A natural way to exploit the sparsity structure of A is to design iterative schemes that involve the coefficient matrix only in the form of matrix-vector products. To achieve this goal, most iterative methods generate iterates x_n by the simple rule
$$x_n = x_0 + q_{n-1}(A)r_0,$$
where $r_0 = f - Ax_0$ denotes the initial residual and q_{n-1} is some polynomial of degree $n-1$. The idea behind such *polynomial based iteration methods* is to choose q_{n-1} such that the scheme converges as fast as possible.

There is one peculiarity which distinguishes this book from the usual literature on polynomial based iteration methods. The conventional approach is based on "matrix manipulations". Here, we emphasize the properties of the underlying polynomials, which will (hopefully) give some more insights into the resulting iteration schemes. Moreover, we exclusively consider (residual) polynomials that are orthogonal with respect to some inner product. There are two main reasons for this choice. First, the three-term recurrence relation of orthogonal polynomials leads to implementations which are cheap in terms of both work and storage. Second, the associated iterates are always characterized by a minimization property.

Two classical examples for polynomial based iteration methods induced by orthogonal polynomials are the conjugate gradient method (cf. Hestenes and Stiefel [81]) and the Chebyshev iteration (cf. Golub and Varga [65]).

It is the purpose of this book to give an overview over the state of the art on polynomial based iteration methods for symmetric systems. However, most of the presented material can be applied to more general situations.

Overview of the Book

In Chapter 1 we introduce notation and some fundamental concepts. In Chapter 2 we concentrate on the theory of orthogonal polynomials. Here, we mainly focus on properties which can be used to our advantage in terms of the corresponding linear solvers. In particular, we emphasize the polynomial minimization problems and discuss in detail the stable generation of recurrence coefficients.

The Chebyshev iteration is based on the *Chebyshev polynomials* with respect to a set containing all eigenvalues of A. In Chapter 3 we outline how to compute the Chebyshev polynomials with respect to the set $E := [a, b] \cup [c, d]$ and show that these polynomials are orthogonal, namely with respect to the *equilibrium distribution* of E. As a by-product we derive a computable expression for the *asymptotic convergence factor* associated with E (cf. Eiermann, Li, and Varga [33]).

Often, one of the key parts of a polynomial based iteration method is the generation of a suitable basis for the associated Krylov space $\mathcal{K}_n(A; r_0)$. In Chapter 4 we show that this task is closely connected to the generation of a basis for the polynomial space Π_{n-1}. In particular we derive the celebrated Lanczos algorithm (cf. Lanczos [87]) from an "orthogonal polynomial point of view".

Chapter 5 is devoted to some "pre-processing" for *parameter dependent schemes*. These methods require some a priori information about the underlying scheme. In this chapter we show how to estimate the spectrum of a given (indefinite) matrix and how to approximate its eigenvalue distribution.

In Chapter 6 we apply the derived results on orthogonal polynomials to the development of so-called *parameter free polynomial based iteration method*. These are methods which require as input only the system matrix A, the right hand side f, and a starting guess x_0. We start by recovering the conjugate gradient method (CG) and the conjugate residual method (CR) (cf. Stiefel [115]). Then we show that these methods may break down when applied to symmetric but indefinite systems. This motivates the search for stable schemes. Again, with the help of orthogonal polynomials we devise schemes which can not break down. They go by names like SYMMLQ, MINRES (cf. Paige and Saunders [95]), MCR (cf. Chandra [18]), and STOD (cf. Stoer and Freund [119]). Finally, we briefly discuss implementations based on normal equations. These are the LSQR method of Paige and Saunders [96] and the CRAIG scheme of Paige [94].

In Chapter 7 we consider parameter dependent methods. In contrast to parameter free methods they need not compute inner products during the iteration. Thus, they are sometimes called *inner product free methods*. We will discuss two particular members of this class. The first one generalizes the classical Chebyshev iteration to indefinite systems. It is well known that the eigenvalue distribution of the given matrix has a great impact on the convergence rate of the various methods. So, in the second approach we start with approximating the eigenvalue distribution and subsequently we compute the orthogonal residual polynomials with respect to this approximation. The generated polynomials may be seen as *weighted Chebyshev polynomials*, where the weight represents the eigenvalue distribution.

In Chapter 8 we introduce and derive one of the "most prominent" sources for symmetric but indefinite linear systems. It stems from an appropriate discretization of the *Stokes equations*. Actually, the variational formulation of these equations may be seen as a saddle point problem. The mathematical treatment of general saddle point problems is also addressed.

Finally, in Chapter 9 we discuss a side problem of polynomial based iteration methods. It is well known that the CG method minimizes the A-norm of the error, which in some applications represents the energy of a certain functional. However, the computation of this quantity involves the solution of the underlying system. In this chapter we devise methods for the approximation of the A-norm of the error, which do not depend on the solution.

Acknowledgments

I am deeply indebted to the many individuals who contributed in various ways to this book. It would fill pages to name them all and probably I would still leave someone unmentioned. I will instead just mention four: Roland Freund, Gene Golub, Jan Modersitzki, and Gerhard Opfer.

I would like to thank Gerhard for his continuous support and guidance over the years. If there is somebody like an academic father and friend, Gerhard truly deserves this title. Gene not only introduced me to the beautiful theory of orthogonal polynomials he also introduced me to the numerical analysis community. Beside this, he taught me what hospitality is really all about. It was always a pleasure to work together with Roland. His insight has had a big influence on this work. Jan, who seems ready to discuss almost anything, read this book, corrected some errors, and made useful suggestions.

I owe thanks to many colleagues for useful comments on the draft version of this book: Wolfgang Hackbusch, Martin Hanke, Marlis Hochbruck, Volker Mehrmann, Alison Ramage, Rolf Rannacher, Lothar Reichel, Henk van der Vorst, and Andy Wathen. David Silvester generously helped me with the section on Stokes problem.

Finally, and foremost, I wish to thank Jutta for her infinite support and her patience over all these years.

I gratefully acknowledge financial support from the DFG (German Research Association).

Contents

1 Introduction   **15**
1.1 Getting Started.. 15
     General Setting 15
     The Startresidual 16
     An Isomorphism 16
     An Approximation Problem 17
     An Inner Product 18
     MATLAB/MATHEMATICA Implementations 18

2 Orthogonal Polynomials   **19**
2.1 General Properties... 19
     Basic Definitions 19
     Three-Term Recurrence Relation 23
     Stieltjes Procedure 26
     Jacobi Matrix 27
     Extremal Property of Orthogonal Polynomials 28
     Zeros of Orthogonal Polynomials 29
     Computing Zeros of Orthogonal Polynomials 30
2.2 Some Applications... 30
     Gaussian Quadrature 30
     Moments and Distribution Functions 34
     Associated Polynomials and Continued Fractions 37
2.3 A Useful Tool 44
     QR Factorization; Tridiagonal Case 44
2.4 Orthogonal Residual Polynomials................................ 48
     Recurrence Relations for Orthogonal Residual Polynomials 48
     Extremal Property of Orthogonal Residual Polynomials 51
2.5 Kernel Polynomials.. 55
     Definition and Orthogonality 55
     Extremal Property of Kernel Polynomials 56

	Recurrence Relations for Kernel Polynomials	60
	Zeros of Kernel Polynomials	65
	Computing Zeros of Kernel Polynomials	69
2.6	Hermite Kernel Polynomials	70
	Definition and Orthogonality	71
	Extremal Property and a Christoffel-Darboux Formula	73
	Recurrence Relations for Hermite Kernel Polynomials	79
	Zeros of Hermite Kernel Polynomials	83
2.7	Orthogonal and (Hermite) Kernel Polynomials	84
	The ME Connection	85
	The MR Connection	87
	The ME - MR Connection	89

3 Chebyshev and Optimal Polynomials — 90

3.1	Basic Definitions	90
	Green's Function	92
	Equilibrium Distribution	94
	Characterization of the Best Approximation	96
3.2	Chebyshev and Optimal Polynomials; One Interval	97
3.3	Chebyshev and Optimal Polynomials; Two Intervals	102
	Basic Observations	102
	One More Extremal Point	103
	Elliptic Functions	110
	A Conformal Mapping	115
	Green's Function for the Union of Two Disjoint Intervals	119
	The Achieser Representation of the Chebyshev Polynomials	121
3.4	Computing an Asymptotic Convergence Factor	125
	The Inverse of the Elliptic Sine	126
	MATLAB Implementation of SNINV	128
	Evaluation of a Theta Function	129
	MATLAB Implementation of ASYMPFAC	130

4 Orthogonal Polynomials and Krylov Subspaces — 132

4.1	Generating a Basis; Orthonormal Case	132
	Lanczos Method	134
	MATLAB Implementation of LANCZOS	134
4.2	Generating a Basis; Monic Case	135

5 Estimating the Spectrum and the Distribution function — 137

5.1	The Model Problem	137
5.2	Estimating the Spectrum	140
5.3	Approximating the Distribution Function	144
	Lanczos Method and Distribution Functions	145
	Monotone Spline	149
	MATLAB Implementation of MPCI	151
	Computing New Orthogonal Polynomials	152

6   Parameter Free Methods   155

6.1	Overview	155
6.2	Implementations Based on Three - Term Recurrences	159
	Basic Algorithm	159
	The CG Approach	161
	MATLAB Implementation of CG	162
	The CR Approach	164
	MATLAB Implementation of CR	165
6.3	CG/CR Applied to Indefinite Systems	166
6.4	Implementations Based on the Monic Basis	173
	The STOD Approach	174
	MATLAB Implementation of STOD	175
	The MCR Approach	176
	MATLAB Implementation of MCR	177
	Modifications	178
6.5	Implementations Based on the Lanczos Basis	179
	The SYMMLQ Approach	180
	MATLAB Implementation of SYMMLQ	183
	The MINRES Approach	185
	MATLAB Implementation of MINRES	187
6.6	Residual Smoothing	188
6.7	A "Non-Feasible" Approach	189
	MATHEMATICA Computation of the Minimal Error	190
6.8	Implementations Based on Normal Equations	191
	The Golub/Kahan Bidiagonalization	192
	QR Factorization; Bidiagonal Case	194
	The LSQR Approach	195
	MATLAB Implementation of LSQR	197
	The CRAIG Approach	199

	MATLAB Implementation of CRAIG	201
6.9	Comparison of the Various Methods	202
	Symmetric Spectrum	207

7 Parameter Dependent Methods 212
7.1 The Chebyshev Iteration for Symmetric Indefinite Systems 212
7.2 Methods Based on the Eigenvalue Distribution 217

8 The Stokes Problem 224
8.1 The Continuous Problem 224
    Hilbert Spaces 225
    The Continuous Stokes Problem 226
    Variational Formulation 227
    Saddle Point Problems 229
    Existence and Uniqueness of Solutions 230
8.2 The Discrete Problem 231
    The Linear System 234
8.3 Some Finite Element Spaces 237

9 Approximating the A-Norm 248
9.1 Energy Norm 248
9.2 Approximating the A-Norm of the Error 250
    CG Case 250
    MATLAB Implementation of cfAerr 256
    Two Examples 256
    General Case 259
    Lower and Upper Bounds 262

10 Bibliography 263

11 Notation 274

12 Index 278

1 Introduction

1.1 Getting Started

In the following sections we introduce some basic notation and discuss some basic principles.

General Setting

We are interested in solving the linear system

$$Ax = f,$$

with a polynomial based iterative method. Here A is a given symmetric $N \times N$ nonsingular matrix with real entries, and $f \in \mathbb{R}^N$ is a given right hand side vector. Given an initial guess $x_0 \in \mathbb{R}^N$ for the solution $x_* := A^{-1}f$, let $r_0 := f - Ax_0$ be the initial residual. A *polynomial based iterative method* is one where one can write the residual $r_n := f - Ax_n$ at step n as

(1.1.1) $$r_n = p_n(A)r_0,$$

where

(1.1.2) $$p_n \in \Pi_n, \quad \text{with} \quad p_n(0) = 1.$$

The polynomial p_n is known as the *residual polynomial*. Due to the interpolatory constraint, it can be written as $p_n(t) = 1 - tq_{n-1}(t)$, where q_{n-1} is some polynomial of degree $n-1$. It is easy to verify that the nth iterate x_n is given by

(1.1.3) $$x_n = x_0 + q_{n-1}(A)r_0.$$

Therefore, the polynomial q_{n-1} is often referred to as the *iteration polynomial*. It is interesting to note that the nth error $\varepsilon_n := x_* - x_n$ can also be written in terms of the residual polynomial

$$\varepsilon_n = A^{-1}r_n = p_n(A)A^{-1}r_0 = p_n(A)\varepsilon_0.$$

It is apparent that $r_n \in \mathcal{K}_n(A; r_0)$ is an element of the nth *Krylov subspace* generated by A and r_0

$$(1.1.4) \qquad \mathcal{K}_n(A; r_0) := \text{span}\{r_0, Ar_0, \ldots, A^{n-1}r_0\}.$$

This is the reason why the iteration (1.1.1) is also called a *Krylov subspace method*.

Any "viable scheme" should deliver a reasonable approximation to the solution after at most L steps, where

$$(1.1.5) \qquad L := \dim \mathcal{K}_N(A, r_0).$$

This termination index depends on r_0 and A. It is the degree of the minimum polynomial of r_0 with respect to A. Moreover, L is known as the *grade* of r_0 with respect to A (cf. Wilkinson [136, p. 37]). Also, L is the smallest integer such that $\mathcal{K}_L(A, r_0)$ is an A-invariant subspace.

The Startresidual

Since any invariant subspace has a basis of eigenvectors z_j, we have

$$(1.1.6) \qquad r_0 = \sum_{j=1}^L \sigma_j z_j, \quad \text{with} \quad z_j^T z_j = 1 \quad \text{and} \quad \sigma_j > 0.$$

Here, we have combined those eigenvectors which belong to the same eigenspace. Let us denote by λ_j the eigenvalues associated with the eigenvectors z_j, i.e.,

$$Az_j = \lambda_j z_j.$$

Since L is minimal, it follows that the λ_j's are distinct. From now on, we assume that they are numbered in increasing order

$$\lambda_1 < \lambda_2 < \cdots < \lambda_L.$$

An Isomorphism

There is an intimate connection between the Krylov subspace $\mathcal{K}_n(A; r_0)$ and the space Π_{n-1} of all polynomials of degree not exceeding $n-1$

$$\mathcal{K}_n(A; r_0) = \{p(A)r_0 : p \in \Pi_{n-1}\}$$

1.1 Getting Started

More precisely, with any element $v \in \mathcal{K}_n(A; r_0)$

$$v = \sum_{j=0}^{n-1} a_j A^j r_0 \quad \leftrightarrow \quad p^{(v)}(t) = \sum_{j=0}^{n-1} a_j t^j$$

we associate the polynomial $p^{(v)} \in \Pi_{n-1}$. Under the assumption that $\dim(\mathcal{K}_n) = n$, i.e., $n \leq L$, the mapping $v \mapsto p^{(v)}$ constitutes an isomorphism between the spaces $\mathcal{K}_n$ and Π_{n-1}.

In the light of this observation, we may define an inner product on $\mathcal{K}_n$ via a given inner product $\langle \cdot, \cdot \rangle$ on Π_{n-1} (or vice versa)

(1.1.7) $$\langle u, v \rangle := \langle p^{(u)}, p^{(v)} \rangle.$$

For convenience we use the same notation for both inner products and the associated norms

(1.1.8) $$\|u\| := \sqrt{\langle u, u \rangle} = \sqrt{\langle p^{(u)}, p^{(u)} \rangle} =: \|p^{(u)}\|.$$

An Approximation Problem

Clearly, the goal for the design of polynomial based iterations is to choose p_n in (1.1.1) such that r_n is small as possible

(1.1.9) $$\|r_n\| = \min\left\{\|p(A)r_0\| : p \in \Pi_n, p(0) = 1\right\},$$

where $\|\cdot\|$ is some given norm on $\mathbb{R}^N$. In view of (1.1.7) and (1.1.8) one may state (1.1.9) in terms of residual polynomials

$$\langle p_n, p_n \rangle = \min\left\{\langle p, p \rangle : p \in \Pi_n, p(0) = 1\right\},$$

So, we are left with the problem of choosing an appropriate inner product $\langle \cdot, \cdot \rangle$ and with the problem of solving the resulting approximation problem. At this point, *orthogonal residual polynomials* ψ_n^{OR} enter into the picture. These are polynomials that satisfy the interpolatory constraint (1.1.2) and that are orthogonal with respect to some given inner product

(1.1.10) $$\langle \psi_k^{\mathrm{OR}}, \psi_l^{\mathrm{OR}} \rangle \begin{cases} > 0 & \text{for } k = l \\ = 0 & \text{for } k \neq l \end{cases}, \quad \text{with } \psi_j^{\mathrm{OR}}(0) = 1.$$

As we will see, it is an intrinsic property of orthogonal polynomials to solve a certain approximation problem. Thus, the ψ_n^{OR}'s are "good candidates" for effective polynomial based iteration methods.

An Inner Product

We will have many occasions to investigate the inner product defined by

(1.1.11) $$\langle p, q \rangle_{\text{GAL}} := r_0^T p(A) q(A) r_0.$$

This particular choice induces the Euclidian inner product for the residuals

(1.1.12) $$r_k^T r_l = r_0^T p_k(A) p_l(A) r_0 = \langle p_k, p_l \rangle_{\text{GAL}}.$$

Consequently, the orthogonality of the residual polynomials ψ_n^{OR}, with respect to $\langle \cdot, \cdot \rangle_{\text{GAL}}$, translates into the following GALerkin condition

$$(r_k^{\text{OR}})^T r_l^{\text{OR}} = 0 \quad \text{for} \quad k \neq l.$$

for the associated residuals $r_n^{\text{OR}} = \psi_n^{\text{OR}}(A) r_0$. Moreover, the eigenvector expansion of the initial residual (1.1.6) leads to the following useful expression for the inner product (1.1.11)

(1.1.13) $$\langle p, q \rangle_{\text{GAL}} = \sum_{j=1}^{L} \sigma_j^2 p(\lambda_j) q(\lambda_j).$$

MATLAB/MATHEMATICA Implementations

In this book we provide for any (interesting) algorithm a MATLAB implementation, except for one case, where we provide a MATHEMATICA implementation. To keep the codes "readable" we did not take advantage of any possible programming trick. Also, we do **not** guarantee that the algorithms are correctly implemented, though we do guarantee that the included performance tests and figures are produced with the supplied programs. If anybody is interested in the codes, just send a nice mail to

na.fischer@na-net.ornl.gov

Also, with all implementations we provide an operation count. The correct counting turns out to be quite a tricky task. After some agonizing we finally decided to count the operations of the provided implementations. In so doing, we are very well aware of the fact that a smart programmer may come up with less expensive implementations. Furthermore, we had to decide on what is a single operation. In accordance with Golub and Van Loan [71, §1.2.4] we define a *flop* to be a single floating point operation. For example, let $x, y \in \mathbb{R}^N$ be vectors and let $\alpha \in \mathbb{R}$ be a scalar, then the operation count for the *saxpy* operation $z = \alpha x + y$ is $2N$ flops.

Finally, none of this would have been fun without the great capabilities and user friendly environment of MATLAB.

2 Orthogonal Polynomials

2.1 General Properties

It is well-known (see, e.g., Szegö [122, Chap. IV]) that the *Chebyshev polynomials of the first kind*

$$(2.1.1) \qquad T_n(t) := \cos(n \arccos(t))$$

are orthogonal with respect to the *continuous* inner product

$$(2.1.2) \qquad \langle p, q \rangle_\omega = \int_{-1}^{1} p(t)q(t)w(t)dt, \quad w(t) = \frac{1}{\sqrt{1-t^2}}.$$

The residual polynomials ψ_n^{OR}, associated with the conjugate gradient method (cf. Example 2.4.8) are orthogonal with respect to the *discrete* inner product (cf. 1.1.13)

$$\langle p, q \rangle_{\text{GAL}} = \sum_{j=1}^{L} \sigma_j^2 p(\lambda_j) q(\lambda_j).$$

In the following sections we will collect the main properties of polynomials that are orthogonal with respect to a continuous or discrete inner product.

Basic Definitions

Let w be a non-negative and (Riemann) integrable *weight function* on the (finite) interval $[a, d]$. Here, $[a, d]$, $a < d$, is a given interval of the real line which may or may not contain the origin. We also assume that $w(t) > 0$ on a sufficiently large subset of $[a, d]$ so that

$$\int_a^d w(t)dt > 0.$$

With w there is associated an inner product

$$(2.1.3) \qquad \langle p, q \rangle_w := \int_a^d p(t)q(t)w(t)dt$$

on the vector space of all polynomials. It is well-known (cf. Chihara [22, §3]) that there exists a system of polynomials that are *orthogonal* with respect to this inner product, i.e., a set of polynomials $\{p_j\}$ such that

$$(2.1.4) \qquad p_j \text{ has exact degree } j \text{ and } \langle p_i, p_j \rangle_w \begin{cases} > 0 & \text{if } i = j, \\ = 0 & \text{if } i \neq j. \end{cases}$$

Obviously, the p_j's are determined by (2.1.4) up to a nonzero factor. Let us specify three conditions which will uniquely determine the orthogonal polynomials. We are mainly concerned with *orthonormal polynomials* $\{\psi_j\}$. These are orthogonal polynomials which have a positive leading coefficient and which are normalized by requiring

$$(2.1.5) \qquad \langle \psi_j, \psi_j \rangle_w = 1.$$

Throughout this book, the set $\{\psi_j\}$ denotes a set of orthonormal polynomials with positive leading coefficient.

Another important set are *monic orthogonal polynomials* $\{\psi_j^{MO}\}$, i.e., orthogonal polynomials with leading coefficient one

$$\psi_j^{MO}(t) = t^j + \ldots$$

Beside orthonormal and monic orthogonal polynomials we will investigate *orthogonal residual polynomials* $\{\psi_j^{OR}\}$ (cf. (1.1.10)), i.e. orthogonal polynomials which fulfill the interpolatory constraint

$$\psi_j^{OR}(0) = 1.$$

We remark that $\psi_j^{OR}(t) = \psi_j(t)/\psi_j(0)$ holds, which implies that (2.1.5) is not achievable for polynomials which have a zero at the origin. It is precisely this restriction which will cause trouble in the design of iterative method for indefinite systems.

In general, the system $\{\psi_j\}$ consists of infinitely many polynomials, but reduces to a finite number L of polynomials $\{\psi_j\}_{j=0}^{L-1}$ for the discrete inner product

$$(2.1.6) \qquad \langle p, q \rangle_L := \sum_{j=1}^{L} p(t_j) q(t_j) w_j^2, \quad w_j \neq 0,$$

2.1 General Properties

where $a \leq t_1 < t_2 < \cdots < t_L \leq d$.

Note that the polynomial

$$(2.1.7) \qquad \psi_L(t) := \prod_{j=1}^{L}(t - t_j)$$

is orthogonal to every other polynomial $\langle \psi_L, q \rangle_L = 0$ and in particular to itself $\langle \psi_L, \psi_L \rangle_L = 0$. Consequently, any orthogonal polynomial of degree L is a multiple of ψ_L. Hence no orthonormal polynomial of degree L with respect to $\langle \cdot, \cdot \rangle_L$ exists. We will, however, use as well the notation ψ_L for orthogonal polynomials of degree L. Also, one may view (2.1.6) as a positive *semidefinite* inner product on the space of all (real) polynomials.

Later on, we will show (see 2.2.4) that orthogonal polynomials with respect to the continuous inner product (2.1.3) are also orthogonal with respect to a related discrete inner product.

It is convenient to rewrite both (2.1.3) and (2.1.6) in terms of *Riemann - Stieltjes integrals* (cf. Smirnov [114, §1], Perron [101, §IV]). We have

$$(2.1.8) \qquad \int_a^d p(t)d\sigma_\omega(t) := \int_a^d p(t)w(t)dt, \quad \text{for } \sigma_\omega(t) = \int_a^t w(x)dx,$$

and

$$(2.1.9) \qquad \int_a^d p(t)d\sigma_L(t) := \sum_{j=1}^{L} p(t_j)w_j^2, \quad \text{for } \sigma_L(t) = \sum_{j=1}^{L} w_j^2 \delta(t - t_j),$$

where δ denotes the Heaviside function

$$\delta(t) := \begin{cases} 1 & t \geq 0, \\ 0 & \text{otherwise.} \end{cases}$$

Notice that $\sigma_L(t)$ has precisely L points of increase, namely the t_j's. So the number of existing orthogonal polynomials equals the number of points of increase of the distribution function.

Another little remark is required here. For any Riemann - Stieltjes integral it holds

$$(2.1.10) \qquad \int_a^d d\sigma(t) = \sigma(d) - \sigma(a).$$

On the other hand, however, for the "discrete case" (2.1.9) we have

$$(2.1.11) \quad \int_a^d d\sigma_L(t) = \sum_{j=1}^L w_j^2 = \sigma_L(d), \quad \text{and} \quad \sigma_L(a) = \begin{cases} 0 & \text{for } a < t_1 \\ w_1^2 & \text{for } a = t_1. \end{cases}$$

So, in view of (2.1.10), the limits of integration should be more properly written as

$$\int_{a-}^d \quad \text{or even} \quad \int_{-\infty}^\infty.$$

For convenience, however, we will use the less complicated notation of (2.1.9).

(2.1.8) and (2.1.9) can be interpreted as a weighted average of p over $[a, d]$. Hence, w is sometimes called *density function*, $\sigma(t)$ is also known as *distribution function*, and $d\sigma(t)$ is called *measure*.

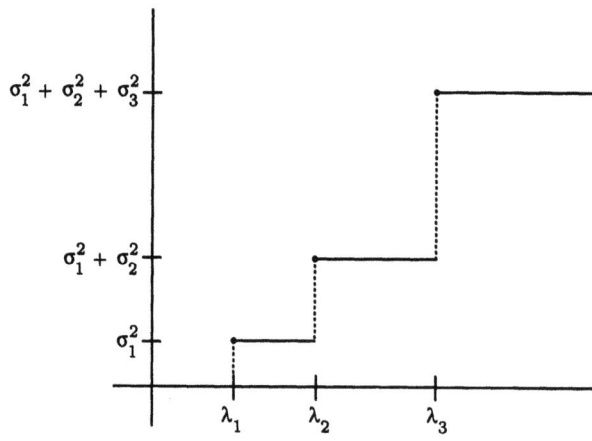

Figure 2.1.1 Distribution function $\sigma(t)$.

The following example illustrates two distribution functions.

Example 2.1.2 (a) The distribution function associated with the Chebyshev weight function $w(t) = 1/\sqrt{1-t^2}$ (cf. (2.1.1)) is

$$(2.1.12) \quad \sigma(t) = \arcsin(t) + \pi/2 \geq 0, \quad \text{since} \quad \sigma'(t) = w(t).$$

2.1 General Properties

(b) The distribution function with respect to the discrete inner product (cf. (1.1.13))

$$\langle p, q \rangle_{\text{GAL}} = \sum_{j=1}^{L} \sigma_j^2 p(\lambda_j) q(\lambda_j) = \int p(t) q(t) d\sigma(t)$$

is given by

$$\sigma(t) = \sum_{j=1}^{L} \sigma_j^2 \delta(t - \lambda_j).$$

It has a jump of height σ_j^2 at any eigenvalue λ_j (compare Figure 2.1.1). □

It will turn out that orthogonal polynomials with respect to continuous inner products (2.1.3) and orthogonal polynomials with respect to discrete inner products (2.1.6) enjoy the same properties. Therefore, in general we will not distinguish between the two inner products $\langle \cdot, \cdot \rangle_\omega$ and $\langle \cdot, \cdot \rangle_L$, respectively. We will use the notation $\langle \cdot, \cdot \rangle$ for both inner products. $\sigma(t)$ will always denote the underlying distribution function. Furthermore, we define

$$\|p\| := \sqrt{\langle p, p \rangle}.$$

Moreover, in the sequel, it is assumed that $n \in \{1, 2, \ldots, L-1\}$, if L is finite, and $n \in \{1, 2, \ldots\}$, if $L = \infty$. This assumption ensures that $\|p\| > 0$ for every polynomial p of degree n which does not vanish identically. The interval $[a, d]$ is therefore called a *supporting set* for $\langle \cdot, \cdot \rangle$.

For convenience we will frequently use one of the following notation

(2.1.13) $\{\psi_j, \langle \cdot, \cdot \rangle; \sigma\}$, $\{\psi_j; \langle \cdot, \cdot \rangle\}$, $\{\psi_j; \sigma\}$, $\{\psi_j, [a, d]; \langle \cdot, \cdot \rangle\}$,

for a set of polynomials orthonormal with respect to an inner product $\langle \cdot, \cdot \rangle$ defined by $\sigma(t)$ on the interval $[a, d]$.

Three-Term Recurrence Relation

One of the most attractive features of orthogonal polynomials is the fact that any three consecutive polynomials are connected by a simple relation.

For example, the (famous) recurrence relationship for the Chebyshev polynomials (2.1.1) is

$$T_0(t) = 1, \quad T_1(t) = t,$$
$$T_j(t) = 2t\, T_{j-1}(t) - T_{j-2}(t), \quad j \geq 2.$$

Let the set $\{p_i\}$ be orthogonal with respect to $\langle \cdot, \cdot \rangle$. Since the p_i's have exact degree i, every tp_{j-1} has a unique expansion of the form

$$(2.1.14) \qquad tp_{j-1}(t) = \sum_{i=0}^{j} a_i^{(j)} p_i(t), \quad a_j^{(j)} > 0.$$

Next, "testing" (2.1.14) with p_i immediately yields the explicit expression for the coefficients

$$a_i^{(j)} = \frac{\langle tp_{j-1}, p_i \rangle}{\langle p_i, p_i \rangle}, \quad i = 0, 1, \ldots, j.$$

Hence, by orthogonality, $a_i^{(j)} = 0$, for $i < j - 2$. We obtain the fundamental three-term recurrence relation

$$(2.1.15) \qquad \gamma_j p_j(t) = (t - \alpha_j) p_{j-1}(t) - \beta_j p_{j-2}(t),$$

where

$$(2.1.16) \qquad \gamma_j = \frac{\langle tp_{j-1}, p_j \rangle}{\langle p_j, p_j \rangle}$$

and

$$(2.1.17) \qquad \alpha_j = \frac{\langle tp_{j-1}, p_{j-1} \rangle}{\langle p_{j-1}, p_{j-1} \rangle}, \quad \beta_j = \frac{\langle tp_{j-1}, p_{j-2} \rangle}{\langle p_{j-2}, p_{j-2} \rangle} = \gamma_{j-1} \frac{\langle p_{j-1}, p_{j-1} \rangle}{\langle p_{j-2}, p_{j-2} \rangle}.$$

The recursion starts with

$$p_{-1}(t) := 0, \quad p_0(t) = p_0.$$

The coefficient β_1 is arbitrary.

For future reference we note that replacing tp_{j-1} in (2.1.14) by some polynomial $p \in \Pi_j$ results in the "Fourier expansion" in terms of orthonormal polynomials

$$(2.1.18) \qquad p(t) = \sum_{i=0}^{j} \langle p, \psi_i \rangle \psi_i(t).$$

With the set $\{p_j\}$ any set $\{k_j p_j\}$, $k_j \neq 0$, is orthogonal with respect to $\langle \cdot, \cdot \rangle$. It is interesting to see what the three-term recurrence coefficients for the

2.1 General Properties

modified polynomials $\hat{p}_j(t) := k_j p_j(t)$ looks like. Working from (2.1.15) we obtain

$$\gamma_j \frac{k_{j-1}}{k_j} \hat{p}_j(t) = (t - \alpha_j)\hat{p}_{j-1}(t) - \beta_j \frac{k_{j-1}}{k_{j-2}} \hat{p}_{j-2}(t),$$

or

$$\hat{\gamma}_j \hat{p}_j(t) = (t - \hat{\alpha}_j)\hat{p}_{j-1}(t) - \hat{\beta}_j \hat{p}_{j-2}(t),$$

with (cf. (2.1.17))

$$\hat{\gamma}_j := \gamma_j \frac{k_{j-1}}{k_j}, \quad \hat{\alpha}_j := \alpha_j,$$

$$\hat{\beta}_j := \beta_j \frac{k_{j-1}}{k_{j-2}} = \gamma_{j-1} \frac{k_{j-2}}{k_{j-1}} \frac{\langle \hat{p}_{j-1}, \hat{p}_{j-1} \rangle}{\langle \hat{p}_{j-2}, \hat{p}_{j-2} \rangle} = \hat{\gamma}_{j-1} \frac{\langle \hat{p}_{j-1}, \hat{p}_{j-1} \rangle}{\langle \hat{p}_{j-2}, \hat{p}_{j-2} \rangle}.$$

Hence, roughly speaking, the choice of the γ_j's determines the nature of the orthogonal polynomials.

A particularly striking example of this statement are monic orthogonal polynomials. Here, the setting $\gamma_j = 1$ "does the job". To be precise, we have

(2.1.19)
$$\psi_{-1}^{MO}(t) := 0, \quad \psi_0^{MO}(t) = 1,$$
$$\psi_j^{MO}(t) = (t - \alpha_j^{MO})\psi_{j-1}^{MO}(t) - \beta_j^{MO}\psi_{j-2}^{MO}(t), \quad j \geq 1,$$

where

(2.1.20) $\alpha_j^{MO} = \dfrac{\langle t\psi_{j-1}^{MO}, \psi_{j-1}^{MO} \rangle}{\langle \psi_{j-1}^{MO}, \psi_{j-1}^{MO} \rangle}$, and $\beta_j^{MO} = \dfrac{\langle \psi_{j-1}^{MO}, \psi_{j-1}^{MO} \rangle}{\langle \psi_{j-2}^{MO}, \psi_{j-2}^{MO} \rangle}$ $(\beta_1^{MO} := 0)$.

Having computed the monic polynomials it is straightforward to devise a recurrence relation for the associated orthonormal polynomials $\psi_j(t) = \psi_j^{MO}(t)/\|\psi_j^{MO}\|$. Here, it holds with appropriate "starting values"

$$\sqrt{\beta_{j+1}^{MO}}\psi_j(t) = (t - \alpha_j^{MO})\psi_{j-1}(t) - \sqrt{\beta_j^{MO}}\psi_{j-2}(t), \quad j \geq 1.$$

In the next subsection we devise a scheme for the direct computation of orthonormal polynomials. Later on, it will turn out that this scheme, applied to $\langle \cdot, \cdot \rangle_{GAL}$, is nothing but the celebrated Lanczos method.

Stieltjes Procedure

The recurrence relation for orthonormal polynomials $\{\psi_j\}$ is special, namely the coefficients satisfy the symmetry relation $\beta_j = \gamma_{j-1}$. The *zero-order moment*

$$(2.1.21) \qquad \nu_0 := \int_a^d d\sigma(t)$$

essentially defines $\psi_0 = \nu_0^{-1/2}$. This leads to the following recurrence relations

$$(2.1.22) \qquad \begin{aligned} &\psi_{-1}(t) := 0, \quad \psi_0(t) = \nu_0^{-1/2}, \\ &\beta_{j+1}\psi_j(t) = (t - \alpha_j)\psi_{j-1}(t) - \beta_j \psi_{j-2}(t), \quad j \geq 1, \end{aligned}$$

where

$$\alpha_j = \langle t\psi_{j-1}, \psi_{j-1}\rangle \quad \text{and} \quad \beta_j = \langle t\psi_{j-2}, \psi_{j-1}\rangle.$$

Notice that β_{j+1} depends on ψ_j. To develop a procedure, alternating recursively between the three-term recurrence relation and the formulae for the coefficients, observe that

$$1 = \langle \psi_j, \psi_j\rangle = \frac{1}{\beta_{j+1}^2}\langle \hat{\psi}_j, \hat{\psi}_j\rangle, \quad \hat{\psi}_j(t) = (t - \alpha_j)\psi_{j-1}(t) - \beta_j\psi_{j-2}(t),$$

and hence

$$(2.1.23) \qquad \beta_{j+1} = \|\hat{\psi}_j\|.$$

Altogether we arrive at the *Stieltjes procedure* for orthonormal polynomials (cf. Gautschi [58]).

Algorithm 2.1.3 *The orthonormal polynomials $\{\psi_j\}_{j=0}^n$ with respect to $\langle \cdot, \cdot \rangle$ may be computed by the following recursive procedure.*

Set
$$\psi_{-1}(t) := 0, \quad \beta_1 := \sqrt{\nu_0}, \quad \psi_0(t) = 1/\beta_1.$$

For $j = 1, 2, \ldots, n$ compute

$$\alpha_j = \langle t\psi_{j-1}, \psi_{j-1}\rangle, \quad \hat{\psi}_j(t) = (t - \alpha_j)\psi_{j-1}(t) - \beta_j\psi_{j-2}(t),$$
$$\beta_{j+1} = \|\hat{\psi}_j\|, \quad \psi_j(t) = \hat{\psi}_j(t)/\beta_{j+1}.$$

2.1 General Properties

To make it a useful tool of computation, it must be supplemented by effective methods for computing the inner products appearing in the scheme. We will discuss this issue whenever we apply the Stieltjes procedure to a particular inner product. The procedure is particularly straightforward if $\langle \cdot, \cdot \rangle$ is a discrete measure, since in this case the inner products are just finite sums.

We remark that Stieltjes' procedure is applicable as long as β_{j+1} does not vanish. For discrete measures this event occurs precisely at step L (compare the comment after (2.1.7))

$$\beta_{L+1} = \|\hat{\psi}_L\| = 0.$$

Jacobi Matrix

An orthogonal polynomial can be seen as the characteristic polynomial of a certain tridiagonal matrix. So, it is not surprising that quite a few results on orthogonal polynomials can be verified with tools from matrix theory. For a collection of such results we refer to Parlett [97, §7].

Let us collect together the three-term recurrence coefficients of $\{\psi_j\}_{j=0}^n$ into an unreduced symmetric tridiagonal matrix J_n, the so-called *Jacobi matrix*,

$$(2.1.24) \qquad J_n := \begin{bmatrix} \alpha_1 & \beta_2 & 0 & \cdots & 0 \\ \beta_2 & \alpha_2 & \ddots & \ddots & \vdots \\ 0 & \ddots & \ddots & \ddots & 0 \\ \vdots & \ddots & \ddots & \ddots & \beta_n \\ 0 & \cdots & 0 & \beta_n & \alpha_n \end{bmatrix}.$$

Furthermore, let us define the $(n+1) \times n$ matrix

$$(2.1.25) \qquad \hat{J}_n := \begin{bmatrix} J_n \\ \beta_{n+1} e_n^T \end{bmatrix}$$

obtained from J_n by appending the row $[0 \ \cdots \ 0 \ \beta_{n+1}]$. Vice versa, J_n can be written in terms of $\hat{J}_n$

$$(2.1.26) \qquad J_n = [I_n \ 0] \hat{J}_n = \hat{J}_n^T \begin{bmatrix} I_n \\ 0 \end{bmatrix}.$$

With the vector

$$\Psi_n(t) := [\,\psi_0(t)\ \ \psi_1(t)\ \ \cdots\ \ \psi_n(t)\,]^T$$

we can rewrite the three-term recurrence relation of the orthonormal polynomials in compact matrix notation as

(2.1.27) $$t\Psi_{n-1}(t) = J_n \Psi_{n-1}(t) + \beta_{n+1}\psi_n(t)e_n$$

or

(2.1.28) $$t\Psi_{n-1}^T(t) = \Psi_{n-1}^T(t) J_n + \beta_{n+1}\psi_n(t)e_n^T = \Psi_n^T(t)\hat{J}_n,$$

respectively.

For example, any residual polynomial $p \in \Pi_n$, $p(0) = 1$, can be written as

$$p(t) = 1 - t\Psi_{n-1}^T(t)y = 1 - \Psi_n^T(t)\hat{J}_n y, \quad y \in \mathbb{R}^n.$$

It is worth noticing that J_n is not necessarily regular. On the other hand, however, $\hat{J}_n$ has always full column rank

(2.1.29) $$\operatorname{rank}\hat{J}_n = n.$$

To see this, recall that all subdiagonal elements $\beta_j = \|\hat{\psi}_{j-1}\|$ of $\hat{J}_n$ are nonzero.

Let us introduce the convenient notation

(2.1.30) $$\{\psi_j, J_n, (\alpha_j, \beta_j), \nu_0; \langle \cdot, \cdot \rangle\}$$

for the orthonormal polynomials ψ_j with respect to $\langle \cdot, \cdot \rangle$ and for associated quantities.

Extremal Property of Orthogonal Polynomials

Among the many properties of the Chebyshev polynomials T_n (cf.(2.1.1)) is the fact that they are uniquely characterized as the solution of an L_2 approximation problem. More precisely, it holds

$$\frac{1}{2^{2n-2}}\int_{-1}^{1} \frac{T_n(t)^2}{\sqrt{1-t^2}}dt \le \int_{-1}^{1} \frac{(t^n - (a_{n-1}t^{n-1} + \cdots + a_0))^2}{\sqrt{1-t^2}}dt$$

2.1 General Properties

for any selection of a_j's.

To solve a least squares approximation problem is an intrinsic property of orthogonal polynomials. The proof of the next theorem is straightforward and can be found in Szegö [122, §3.1].

Theorem 2.1.4 Let $p_n \in \Pi_n$ be a monic polynomial. Then, the following conditions are equivalent:

(a) $\langle p_n, q \rangle = 0$, for all $q \in \Pi_{n-1}$;

(b) $\|p_n\| = \min \{ \|t^n - q(t)\| : q \in \Pi_{n-1} \}$.

Zeros of Orthogonal Polynomials

The zeros of orthogonal polynomials exhibit certain regularities which will be listed in the next theorem (see, e.g., Szegö [122, §3.3]).

Theorem 2.1.5 Let $\{\psi_j, [a,d]; \sigma\}$ (cf. (2.1.13)) be given. Moreover, let $\theta_k^{(j)}$, $k = 1, 2, \ldots, j$, denote the zeros of ψ_j.

(a) The zeros of ψ_n are all real, simple and are located in (a,d)

$$a < \theta_1^{(n)} < \theta_2^{(n)} < \cdots < \theta_n^{(n)} < d.$$

(b) The zeros of ψ_n and ψ_{n+1} separate each other

$$\theta_1^{(n+1)} < \theta_1^{(n)} < \theta_2^{(n+1)} \cdots < \theta_n^{(n)} < \theta_{n+1}^{(n+1)}.$$

(c) Between two zeros of ψ_m there is at least one zero of ψ_n, $n > m$.

(d) In an open interval in which $\sigma(t)$ is constant, ψ_n has at most one zero.

Computing Zeros of Orthogonal Polynomials

As a direct consequence of (2.1.27) we obtain the fundamental fact that an orthogonal polynomial is the characteristic polynomial of "its Jacobi matrix" (compare Golub and Welsch [66]).

> **Theorem 2.1.6** Let $\{\psi_n, J_n; \sigma\}$ (cf. (2.1.30)) be given. Then, any zero $\theta_j^{(n)}$ of ψ_n is an eigenvalue of J_n with eigenvector $\Psi_{n-1}(\theta_j^{(n)})$.

So, to compute the zeros of orthogonal polynomials, set up the Jacobi matrices and find their eigenvalues.

2.2 Some Applications

Orthogonal polynomials have been used extensively in various branches of numerical analysis. In the following section we list some applications which are useful in connection with polynomial iteration methods.

Gaussian Quadrature

In this section we show that any set of orthogonal polynomials is also orthogonal with respect to a discrete inner product. Here, the key approach is the investigation of the associated Gaussian quadrature formula which will then induce the wanted inner product.

A beautiful survey of Gaussian quadrature and related topics, in particular from a historical point of view, can be found in Gautschi [58].

The Gaussian quadrature rule has maximum algebraic degree of exactness: there exist weights $\{\tau_j^{(n)}\}_{j=1}^n$ and knots $\{\theta_j^{(n)}\}_{j=1}^n$, the zeros of ψ_n, such that for every polynomial p of degree at most $2n - 1$

$$(2.2.1) \qquad \int_a^d p(t)d\sigma(t) = \sum_{j=1}^n \left(\tau_j^{(n)}\right)^2 p(\theta_j^{(n)}).$$

2.2 Some Applications

The weights $\tau_j^{(n)}$ are called *Christoffel numbers*. They can be expressed in various ways. Usually, they are given in terms of the associated orthonormal polynomials (cf. Szegö [122, §3.4])

(2.2.2) $$\left(\tau_j^{(n)}\right)^{-2} = \sum_{i=0}^{n-1} \psi_i^2(\theta_j^{(n)}) = \Psi_{n-1}^T\left(\theta_j^{(n)}\right)\Psi_{n-1}\left(\theta_j^{(n)}\right).$$

Hence, in view of Theorem 2.1.6, formula (2.2.1) is determined by the eigenvalues and eigenvectors of J_n. For computational purposes it is useful to note that from (2.2.2) the normalized eigenvectors $s_j^{(n)}$ look like

$$1 = \left(\tau_j^{(n)}\Psi_{n-1}\left(\theta_j^{(n)}\right)\right)^T\left(\tau_j^{(n)}\Psi_{n-1}\left(\theta_j^{(n)}\right)\right) =: \left(s_j^{(n)}\right)^T s_j^{(n)}.$$

In particular, by (2.1.22) and (2.1.21), we obtain for the first component

$$e_1^T s_j^{(n)} = \tau_j^{(n)}\psi_0 = \tau_j^{(n)}/\sqrt{\nu_0}$$

and finally (compare Wilf [135, Ch. 2])

(2.2.3) $$\tau_j^{(n)} = \sqrt{\nu_0}\, e_1^T s_j^{(n)}.$$

That is, we only have to compute the eigenvalues and the first component of the normalized eigenvectors. In general, these quantities are calculated by special versions of the QR-algorithm (cf. Golub and Welsch [66]).

Gaussian quadrature with respect to a discrete distribution function σ_L (cf. (2.1.9)) is special. Here, we have

$$\sum_{j=1}^L w_j^2 f(t_j) = \int_a^d f(t)d\sigma_L(t) = \sum_{j=1}^L \left(\tau_j^{(L)}\right)^2 f(\theta_j^{(L)})$$

for any (smooth) function f. To justify this, first recall from (2.1.7) that the zeros of ψ_L coincide with the abscissas $\theta_j^{(L)} = t_j$. Second, let us show that the square of the Christoffel number $\tau_j^{(L)}$ is just the weight w_j^2. To this end, let $\{l_i\}_{i=1}^L$ denote the Lagrange basis functions with respect to $t_1 < t_2 < \ldots < t_L$, i.e.,

$$l_i(t_j) = \begin{cases} 1 & \text{for } i = j \\ 0 & \text{for } i \neq j. \end{cases}$$

Then, we have in view of (2.2.1)

$$w_i^2 = \sum_{j=1}^L w_j^2 l_i(t_j) = \sum_{j=1}^L \left(\tau_j^{(L)}\right)^2 l_i(t_j) = \left(\tau_i^{(L)}\right)^2.$$

We are mainly interested in evaluating integrals involving polynomials $\langle p, 1 \rangle$. Here one can avoid the computation of eigenvalues and eigenvectors (cf. Fischer and Golub [40]). Since J_n is Hermitian, there exists a unitary matrix U_n with

$$U_n^T J_n U_n = \operatorname{diag}(\theta_1^{(n)}, \theta_2^{(n)}, \ldots, \theta_n^{(n)}) \ (=: \Theta_n),$$

where each column of U_n is a normalized eigenvector of J_n. Therefore, we have by (2.2.1) and (2.2.3)

$$\langle p, 1 \rangle = \nu_0 \sum_{j=1}^n \left(e_1^T s_j^{(n)}\right)^2 p(\theta_j^{(n)})$$

$$= \nu_0 \sum_{j=1}^n e_1^T U_n^T p(\Theta_n) U_n e_1$$

$$= \nu_0 e_1^T p(J_n) e_1.$$

This is probably the most compact version of the Gaussian quadrature formula.

Theorem 2.2.1 *Let $\{J_n, \nu_0; \langle \cdot, \cdot \rangle\}$ (cf. (2.1.30)) be given. Then, for every polynomial p of degree at most $2n - 1$*

$$\langle p, 1 \rangle = \nu_0 e_1^T p(J_n) e_1.$$

A similar application can be made to rational functions, provided that J_n is nonsingular. We only state a special case.

2.2 Some Applications

Theorem 2.2.2 *Let $\{J_n, \nu_0; \langle \cdot, \cdot \rangle\}$ (cf. (2.1.30)) be given.*

(a) If J_n is nonsingular, then the Gaussian quadrature formula for the function t^{-1} reads

$$\langle t^{-1}, 1 \rangle \sim \nu_0 e_1^T J_n^{-1} e_1.$$

(b) If J_L is nonsingular, then (a) is exact for the discrete inner product (2.1.6)

$$\langle t^{-1}, 1 \rangle_L = \nu_0 e_1^T J_L^{-1} e_1.$$

We are now in a position to show that polynomials which are orthogonal with respect to a continuous inner product are also orthogonal with respect to a discrete inner product. Suppose that $q \in \Pi_{n-1}$, then

$$\begin{aligned}
0 = \langle \psi_n, q \rangle &= \nu_0 e_1^T \psi_n(J_n) q(J_n) e_1 \\
&= \sum_{j=1}^{n} \left(\tau_j^{(n)} \right)^2 \psi_n(\theta_j^{(n)}) q(\theta_j^{(n)}) \\
&=: \int_a^d \psi_n(t) q(t) d\tau(t) \\
&=: \langle \psi_n, q \rangle_n,
\end{aligned}$$
(2.2.4)

i.e., ψ_n is orthonormal relative to $\langle \cdot, \cdot \rangle_n$. The distribution function

(2.2.5)
$$\tau(t) = \sum_{j=1}^{n} \left(\tau_j^{(n)} \right)^2 \delta(t - \theta_j^{(n)})$$

is determined by the knots and weights of the associated Gaussian quadrature rule.

The "new" inner product $\langle \cdot, \cdot \rangle_n$ may be viewed as an approximation to the given inner product $\langle \cdot, \cdot \rangle$. Also, the "low order" distribution function $\tau(t)$ has to be close to the given distribution function $\sigma(t)$. A special aspect of this connection will be discussed in the next subsection.

Moments and Distribution Functions

We start with an example.

Example 2.2.3 We investigate, once more, the Chebyshev polynomials $T_n(t) = \cos(n \arccos(t))$. The associated distribution function is given by (cf. (2.1.12))

$$\sigma(t) = \arcsin(t) + \pi/2.$$

Furthermore, the corresponding orthonormal polynomials ψ_n look like

$$\psi_n(t) = \begin{cases} \sqrt{\frac{1}{\pi}} T_0(t) & \text{for } n = 0, \\ \sqrt{\frac{2}{\pi}} T_n(t) & \text{for } n > 0. \end{cases}$$

The zeros of T_n are

$$\theta_{j+1}^{(n)} = \cos\left(\frac{2j+1}{2n}\pi\right), \quad j = 0, 1, \ldots, n-1.$$

It is easy to verify that the Christoffel numbers (2.2.2)

$$\begin{aligned}
\left(\tau_{j+1}^{(n)}\right)^{-2} &= \sum_{k=0}^{n-1} \psi_k^2\left(\theta_{j+1}^{(n)}\right) \\
&= -\frac{1}{\pi} + \frac{2}{\pi}\sum_{k=0}^{n-1} \cos\left(k\frac{2j+1}{2n}\pi\right)^2 \\
&= \frac{n}{\pi} + \frac{1}{2}\sum_{k=0}^{n} \cos\left(k\frac{2j+1}{n}\pi\right) \\
&= \frac{n}{\pi}
\end{aligned}$$

are independent of j and, consequently, the "low order" distribution function (2.2.5) reduces to

$$(2.2.6) \qquad \tau(t) = \frac{\pi}{n}\sum_{j=1}^{n} \delta(t - \theta_j^{(n)}).$$

It is worth noticing, that the Chebyshev weight function (2.1.2) is the only weight function (up to a linear transformation) for which the Gaussian quadrature formula has equal weights (cf. Posse [103]).

2.2 Some Applications

So, in view of (2.2.4), we recover the well-known trigonometric identity

$$0 = \langle T_k, T_l \rangle_n = \frac{\pi}{n} \sum_{j=1}^{n} T_k(\theta_j^{(n)}) T_l(\theta_j^{(n)})$$

$$= \frac{\pi}{n} \sum_{j=1}^{n} \cos\left(k \frac{2j-1}{2n} \pi\right) \cos\left(l \frac{2j-1}{2n} \pi\right),$$

where $0 \leq l < k \leq n$.

The next figure shows the two distribution functions $\sigma(t)$ (cf. (2.1.12)) and $\tau(t)$ (cf. (2.2.6)) for the case $n = 5$.

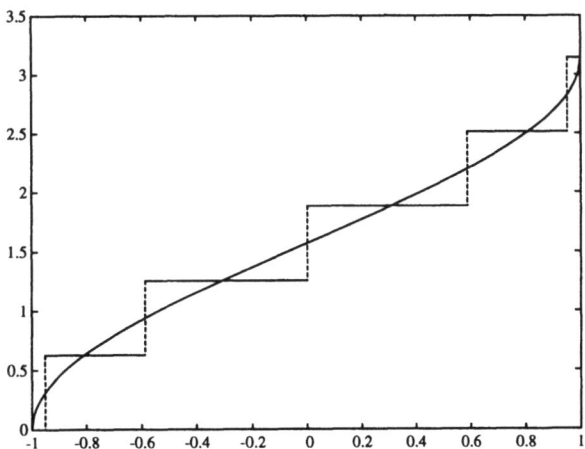

Figure 2.2.4 Distribution functions $\sigma(t)$ and $\tau(t)$ for $n = 5$.

Note that the horizontal steps and the vertical steps of $\tau(t)$ intersect $\sigma(t)$ exactly 9 times in $[-1, 1]$. In other words, the difference function $\tau(t) - \sigma(t)$ has 9 sign changes in (-1,1).

There is yet another fact we would like to point out. Since $\cos(t) = \sin(t + \pi/2)$, the arcsin function intersects $\tau(t)$ precisely at the "half steps" $\pi(j - 1/2)/n$. □

It is subject of this section to show that the "interlocking property", illustrated in Figure 2.2.4, is valid for any given distribution function and its approximation generated by the Gaussian quadrature rule.

We just learned from Theorem 2.2.1 and (2.2.4) that, for all $p \in \Pi_{2n-1}$

$$(2.2.7) \qquad \int_a^d p(t)d\sigma(t) = \langle p, 1 \rangle = \langle p, 1 \rangle_n = \int_a^d p(t)d\tau(t).$$

In other words, the first $2n$ *modified moments*

$$\nu_j = \nu(p_j; \sigma) := \int_a^d p_j(t)d\sigma(t), \quad j = 0, 1, \ldots, 2n-1,$$

of $\langle \cdot, \cdot \rangle$ and $\langle \cdot, \cdot \rangle_n$ coincide. Here, $\{p_j\}_{j=0}^{2n-1}$ is a given set of polynomials of exact degree j.

Vice versa, it is well-known that the set $\{\nu(p_j; \sigma)\}_{j=0}^{2n-1}$ already determines the orthogonal polynomials up to degree $n-1$ with respect to $\sigma(t)$. In fact it is possible to use the moments ν_j to build up recursively the wanted three-term recurrence coefficients. The resulting algorithm, called *modified Chebyshev algorithm*, only requires $O(n^2)$ arithmetic operations. In particular, beside the computation of the moments, no further evaluation of inner products is necessary (see, e.g., Sack and Donavan [109], Wheeler [132], Gautschi [60], and Fischer and Golub [40]).

Thus the modified Chebyshev algorithm is (in general) preferable on account of its superior efficiency to the Stieltjes algorithm. Unfortunately, there are instances where the Chebyshev algorithm becomes "highly" unstable. The factors responsible for the stability properties of the scheme are discussed, for example, in Gautschi [61].

Condition (2.2.7) implies, roughly speaking, that the distribution function $\tau(t)$ (cf. (2.2.5)) has to be close to the distribution function $\sigma(t)$, defined in (2.1.8) or (2.1.9), respectively. This statement can be made more precise, in the sense of the following theorem due to Karlin and Shapley [85, Theorem 22.2].

Theorem 2.2.5 *Let the distribution function $\tau(t)$ and $\sigma(t)$ have the same modified moments up to degree $2n-1$. Then, the difference function $\rho(t) := \tau(t) - \sigma(t)$ has precisely $2n-1$ sign changes in $[a, d]$.*

2.2 Some Applications

Proof. We offer the following proof for completeness. We start by noting that (2.2.7) in particular implies $\langle 1, 1 \rangle = \langle 1, 1 \rangle_n$, or equivalently (cf. (2.1.11))

(2.2.8) $$\rho(d) = \tau(d) - \sigma(d) = 0 \ (= \rho(a-)).$$

Now assume, to the contrary, that $\rho(t)$ has fewer than $2n - 1$ sign changes. Then there exists a polynomial $p \in \Pi_{2n-2}$ with

(2.2.9) $$\int_a^d p(t)\rho(t)dt > 0.$$

On the other hand, integration by parts leads to

(2.2.10) $$\int_a^d p(t)\rho(t)dt = q(t)\rho(t)\Big|_{a-}^d - \int_a^d q(t)(d\tau(t) - d\sigma(t)),$$

where $q \in \Pi_{2n-1}$ is defined by $q' = p$. But from (2.2.8) and (2.2.7) we conclude that the right hand side of (2.2.10) vanishes in contradiction to (2.2.9).

Finally, a monotonicity argument shows that $\rho(t)$ can have not more than $2n - 1$ sign changes. □

We note that the statement of Theorem 2.2.5 appears implicitly in the second proof of the Separation Theorem 3.41.1 in Szegö [122]. This proof, however, goes back to Stieltjes[117].

Associated Polynomials and Continued Fractions

We start with a "motivating example". The CG method (cf. Example 2.4.8)

$$\varepsilon_n^{OR} = x_* - x_n^{OR} = \psi_n^{OR}(A)\varepsilon_0$$

minimizes the A-norm of the error

$$\|\varepsilon_n^{OR}\|_A^2 = (\varepsilon_n^{OR})^T A \varepsilon_n^{OR} = (r_n^{OR})^T A^{-1} r_n^{OR}$$

with respect to the Krylov subspace $\mathcal{K}_n(A; r_0)$. The direct computation of $\|\varepsilon_n^{OR}\|_A$ involves the solve of a linear system with coefficient matrix A. It is subject of this section to come up with a procedure which cheaply approximates the quantity $\|\varepsilon_n^{OR}\|_A$. The scheme is based on the following observation. In view of (1.1.11), one may express

$$\|\varepsilon_n^{OR}\|_A^2 = (r_0 \psi_n^{OR}(A))^T A^{-1} \psi_n^{OR}(A) r_0$$
$$= \langle \psi_n^{OR}, t^{-1} \psi_n^{OR} \rangle_{GAL}$$
$$= \int \frac{(\psi_n^{OR}(t))^2}{t} d\sigma(t)$$

in terms of an integral with respect to the (unknown) distribution function $\sigma(t)$. The idea is to approximate this integral by means of Gaussian quadrature. This approach, however, involves the knowledge of a set of orthogonal polynomials with respect to either $\langle \cdot, \cdot \rangle_{GAL}$ or $\langle \cdot, t^{-1} \cdot \rangle_{GAL}$, and the knowledge of the associated zero-order moment(s). One should keep in mind (cf. Example 2.4.8) that the CG residual polynomials are just the orthogonal polynomials with respect to $\langle \cdot, \cdot \rangle_{GAL}$.

We treat a more general case. That is, we assume that σ is a given distribution function and ξ is a given point on the real axis. Then, we discuss the approximation of

(2.2.11) $$F(\xi) := \int_a^d \frac{d\sigma(t)}{\xi - t} = \langle (\xi - t)^{-1}, 1 \rangle$$

under the assumption that the recurrence coefficients of the monic orthogonal polynomials ψ_n^{MO} with respect to σ are given. To be precise, we have (compare (2.1.19))

(2.2.12) $$\psi_j^{MO}(t) = (t - \alpha_j) \psi_{j-1}^{MO}(t) - \beta_j \psi_{j-2}^{MO}(t), \quad j \geq 1,$$

with "starting values"

$$\psi_{-1}^{MO}(t) := 0 \quad \text{and} \quad \psi_0^{MO}(t) = 1.$$

Here, the coefficient β_1 is arbitrary, but will be set equal to ν_0

(2.2.13) $$\beta_1 := \nu_0 = \int_a^d d\sigma(t).$$

It is important to note, that the β_j's, $j \leq L$, do not vanish. This is a direct consequence of (2.1.17).

2.2 Some Applications

The expression (2.2.11) may be viewed as the zero-order modified moment $\tilde{\nu}_0$, where

(2.2.14) $$\tilde{\nu}_j(\xi) := \int_a^d \frac{\psi_j^{MO}(t)}{\xi - t} d\sigma(t), \quad j \geq 0.$$

The (stable) computation of these moments is also subject of this section. Note, that the knowledge of these moments would enable us to compute the orthogonal polynomials with respect to $\langle (\xi - t)^{-1} \cdot, \cdot \rangle$.

The results rely on unexpected relations between continued fractions and orthogonal polynomials.

To begin with, consider the polynomials

(2.2.15) $$\tilde{\psi}_n(t) := \int_a^d \frac{\psi_n^{MO}(t) - \psi_n^{MO}(x)}{t - x} d\sigma(x),$$

which are called the polynomials *associated* with the orthogonal polynomials ψ_n^{MO} or the *polynomials of the second kind*. Clearly, $\tilde{\psi}_n$ has degree $n-1$. Surprisingly enough, these polynomials satisfy the recurrence relation (2.2.12)

(2.2.16) $$\tilde{\psi}_j(t) = (t - \alpha_j)\tilde{\psi}_{j-1}(t) - \beta_j \tilde{\psi}_{j-2}(t), \quad j \geq 1,$$

but with different starting values

$$\tilde{\psi}_{-1}(t) := -1 \quad \text{and} \quad \tilde{\psi}_0(t) := 0.$$

Notice that the choice (2.2.13) implies $\tilde{\psi}_1(t) = \nu_0$. To justify (2.2.16), apply (2.2.12)

$$\psi_n^{MO}(t) - \psi_n^{MO}(x) = (t-x)\psi_{n-1}^{MO}(x) + (t - \alpha_n)\big(\psi_{n-1}^{MO}(t) \\ - \psi_{n-1}^{MO}(x)\big) - \beta_n\big(\psi_{n-2}^{MO}(t) - \psi_{n-2}^{MO}(x)\big)$$

divide both sides by $t - x$ and finally analyze

$$\tilde{\psi}_n(t) = \int_a^d \psi_{n-1}^{MO}(x) d\sigma(x) + (t - \alpha_n)\tilde{\psi}_{n-1}(t) - \beta_n \tilde{\psi}_{n-2}(t).$$

Observe that, by orthogonality, the integral on the right vanishes for $n \geq 2$.

We just showed that both $\{\psi_n^{MO}(\xi)\}$ and $\{\tilde{\psi}_n(\xi)\}$ satisfy, for a fixed parameter ξ, the homogeneous second-order recurrence relation

(2.2.17) $$y_j = (\xi - \alpha_j)y_{j-1} - \beta_j y_{j-2}, \quad j \geq 1.$$

Moreover, since the so-called *Casorati determinant*

$$\begin{vmatrix} \psi_0^{MO}(\xi) & \tilde{\psi}_0(\xi) \\ \psi_1^{MO}(\xi) & \tilde{\psi}_1(\xi) \end{vmatrix} = \begin{vmatrix} 1 & 0 \\ \xi - \alpha_1 & \nu_0 \end{vmatrix} = \nu_0 > 0$$

does not vanish at $j = 0, 1$, we see that $\{\psi_n^{MO}(\xi)\}$ and $\{\tilde{\psi}_n(\xi)\}$ form a fundamental set, i.e., any solution of (2.2.17)

(2.2.18) $$y_j = a\psi_j^{MO}(\xi) + b\tilde{\psi}_j(\xi), \quad j \geq 1,$$

can be expressed as a linear combination of these quantities.

A straightforward induction argument (see, e.g., Cheney [21, Chap. 5, Theorem 1]) shows that the quotient $\tilde{\psi}_n(\xi)/\psi_n^{MO}(\xi)$ is the nth convergent of a continued fraction

(2.2.19)
$$\frac{\tilde{\psi}_n(\xi)}{\psi_n^{MO}(\xi)} = \cfrac{\beta_1}{\xi - \alpha_1 - \cfrac{\beta_2}{\xi - \alpha_2 - \cfrac{\beta_3}{\ddots\cfrac{}{\xi - \alpha_{n-1} - \cfrac{\beta_n}{\xi - \alpha_n}}}}}$$

$$=: \frac{\beta_1}{\xi - \alpha_1 -} \frac{\beta_2}{\xi - \alpha_2 -} \cdots \frac{\beta_n}{-\xi - \alpha_n}.$$

On the other hand, since $\tilde{\psi}_n(\xi)$ is defined as integral of a polynomial, we may evaluate the left hand side of (2.2.19) by means of the Gaussian quadrature rule with respect to $\sigma(t)$

(2.2.20)
$$\frac{\tilde{\psi}_n(\xi)}{\psi_n^{MO}(\xi)} = \frac{1}{\psi_n^{MO}(\xi)} \int_a^d \frac{\psi_n^{MO}(\xi) - \psi_n^{MO}(t)}{\xi - t} d\sigma(t)$$

$$= \sum_{j=1}^n \frac{(\tau_j^{(n)})^2}{\xi - \theta_j^{(n)}} \left(1 - \frac{\psi_n^{MO}(\theta_j^{(n)})}{\psi_n^{MO}(\xi)}\right)$$

$$= \nu_0 e_1^T (\xi I_n - J_n)^{-1} e_1,$$

2.2 Some Applications

where the $\theta_j^{(n)}$ denote the zeros of ψ_n^{MO} and J_n is the Jacobi matrix associated with σ. Of course, the derivations above assume that ξ is not a zero of ψ_n^{MO}.

We conclude that the quotient $\tilde{\psi}_n(\xi)/\psi_n^{\mathrm{MO}}(\xi)$ and the continued fraction (2.2.19) provide alternative expressions for the n-point Gaussian quadrature formula applied to the integral $F(\xi)$ (cf. (2.2.11)).

We are now ready to state the announced theorem. Part (a) is a direct consequence of (2.2.19), (2.2.20), and Theorem 2.2.2(b). Part (b) is not as easy to verify. A proof can be found, for example, in Perron [101, Satz 4.1, Satz 4.2].

Theorem 2.2.6

(a) Let $\langle \cdot, \cdot \rangle_L$ (cf. (2.1.6)) denote an inner product with precisely L points of increase. Moreover, let $\{\psi_j^{\mathrm{MO}}, \tilde{\psi}_j, (\alpha_j, \beta_j); \langle \cdot, \cdot \rangle_L\}$ (cf. (2.1.13)) be given. If $\psi_L^{\mathrm{MO}}(\xi) \neq 0$ does not vanish at the given point ξ, then

$$\frac{\tilde{\psi}_L(\xi)}{\psi_L^{\mathrm{MO}}(\xi)} = \frac{\beta_1}{\xi - \alpha_1 -} \frac{\beta_2}{\xi - \alpha_2 -} \cdots \frac{\beta_L}{-\xi - \alpha_L} = \langle (\xi - t)^{-1}, 1 \rangle_L.$$

(b) Let $\langle \cdot, \cdot \rangle_\omega$ (cf. (2.1.3)) denote an inner product with infinitely points of increase. Moreover, let $\{\psi_j^{\mathrm{MO}}, \tilde{\psi}_j; \langle \cdot, \cdot \rangle_\omega\}$ (cf. (2.1.13)) be given. If $\xi \notin [a, d]$ does not belong to the supporting set of $\langle \cdot, \cdot \rangle_\omega$, then

$$\lim_{n \to \infty} \frac{\tilde{\psi}_n(\xi)}{\psi_n^{\mathrm{MO}}(\xi)} = \langle (\xi - t)^{-1}, 1 \rangle_\omega.$$

There is yet another application of Theorem 2.2.6(b). It is concerned with the construction of orthogonal polynomials with respect to the modified distribution

$$d\tilde{\sigma}(t) := \frac{d\sigma(t)}{\xi - t}.$$

According to Section 2.2, this task may be accomplished by the modified Chebyshev algorithm, provided some modified moments, e.g., the one defined in (2.2.14), are known. It is easy to verify that $\tilde{\nu}_j(\xi)$ is a linear combination of $\psi_j^{MO}(\xi)$ and $\tilde{\psi}_j(\xi)$, namely

$$(2.2.21) \qquad \tilde{\nu}_j(\xi) = F(\xi)\psi_j^{MO}(\xi) - \tilde{\psi}_j(\xi).$$

Consequently, $\{\tilde{\nu}_n(\xi)\}$ is another solution of (2.2.17)

$$(2.2.22) \qquad \tilde{\nu}_j(\xi) = (\xi - \alpha_j)\tilde{\nu}_{j-1}(\xi) - \beta_j\tilde{\nu}_{j-2}(\xi), \quad j \geq 1,$$

with starting values

$$\tilde{\nu}_{-1}(\xi) := 1 \quad \text{and} \quad \tilde{\nu}_0(\xi) = F(\xi).$$

On the first glance, this seems to be the perfect recursion for determining the wanted modified moments. However, great care has to be taken in any direct implementation of (2.2.22). The reason is that $\{\tilde{\nu}_n(\xi)\}$ constitutes a *minimal solution* of (2.2.17), i.e., there exists a linearly independent solution $\{p_n(\xi)\}$ of (2.2.17) with

$$(2.2.23) \qquad \lim_{n \to \infty} \frac{\tilde{\nu}_n(\xi)}{p_n(\xi)} = 0.$$

To justify this, observe that by Theorem 2.2.6(b) and (2.2.21) the set of monic polynomials $\{\psi_n^{MO}(\xi)\}$ "does the job"

$$(2.2.24) \qquad \frac{\tilde{\nu}_n(\xi)}{\psi_n^{MO}(\xi)} = F(\xi) - \frac{\tilde{\psi}_n(\xi)}{\psi_n^{MO}(\xi)} \to 0.$$

In fact, it is easily seen that (2.2.23) holds for any solution (2.2.18) of (2.2.17) which is not proportional to $\tilde{\nu}_n(\xi)$. Moreover, observe that the minimal solution is, up to a constant multiple, unique. Hence, the single starting value $\tilde{\nu}_{-1}$ already determines the whole sequence.

Suppose we generate by (2.2.17) a set $\{y_n\}$ with starting values $y_0 = \tilde{\nu}_0(\xi) + \varepsilon$ and $y_1 = \tilde{\nu}_1(\xi) + \varepsilon$, where ε is some small number due to roundoff errors, for example. Then, in general, $\{y_n\}$ will be independent of $\{\tilde{\nu}_n(\xi)\}$ and hence

$$\left|\frac{y_n - \tilde{\nu}_n(\xi)}{\tilde{\nu}_n(\xi)}\right| \to \infty.$$

Hence, any roundoff error in the start phase of the recurrence (2.2.22), e.g., in the computation of $F(\xi)$, will eventually "blow up" the scheme.

2.2 Some Applications

We summarize (for further details see Gautschi [56], [57] and Wimp[137, §2.6]):

> **Theorem 2.2.7** *The recurrence (2.2.22) is unstable for the minimal solution $\{\tilde{\nu}_n(\xi)\}$.*

Based on continued fractions, Gautschi [56] devised an effective and stable algorithm for the computation of the modified moments (2.2.14) We will make use of this approach in Section 9.2.

We finish this section with a very useful and elegant characterization of orthogonal polynomials and the corresponding associated polynomials.

In (2.2.24) we were interested in case $n \to \infty$. Now we are concerned with the asymptotic of the case $|\xi| \to \infty$. For the derivation of (2.2.24) we used (implicitly) the expression

$$(2.2.25) \qquad \psi_n^{\text{MO}}(\xi) F(\xi) - \tilde{\psi}_n(\xi) = \tilde{\nu}_n(\xi).$$

Let us have a closer look at the right hand side. The orthogonality of ψ_n^{MO} combined with (2.2.14) implies

$$0 = \int_a^d \psi_n^{\text{MO}}(t) \frac{\xi^n - t^n}{\xi - t} d\sigma(t) = \xi^n \tilde{\nu}_n(\xi) - \int_a^d \psi_n^{\text{MO}}(t) \frac{t^n}{\xi - t} d\sigma(t)$$

and hence we find that as $|\xi| \to \infty$

$$\tilde{\nu}_n(\xi) = O\left(\frac{1}{\xi^{n+1}}\right).$$

This together with (2.2.25) proves one direction of the following theorem. The other direction is straightforward to verify and can be found, for example, in Achieser [3, pp. 157].

> **Theorem 2.2.8** Let $\{\langle \cdot, \cdot \rangle; \sigma\}$ (cf. (2.1.13)) be given. Furthermore let F be defined as in (2.2.11). Then, ψ_n is an orthogonal polynomial with respect to $\langle \cdot, \cdot \rangle$ and $\tilde{\psi}_n$ is the corresponding associated polynomial if, and only if
>
> $$\psi_n(\xi)F(\xi) - \tilde{\psi}_n(\xi) = O\left(\frac{1}{\xi^{n+1}}\right), \quad \text{for } |\xi| \to \infty.$$

2.3 A Useful Tool

In the next sections we will have occasion to solve linear systems involving the Jacobi matrix J_n or the compound matrix $\hat{J}_n^T \hat{J}_n$ (cf. (2.1.24) and (2.1.25)). A crucial step towards effective schemes is the utilization of the QR factorization of $\hat{J}_n$.

QR Factorization; Tridiagonal Case

In this section we will briefly outline how to compute an $(n+1) \times (n+1)$ unitary matrix Q_n and an $n \times n$ upper triangular matrix

$$(2.3.1) \qquad R_n = \begin{bmatrix} r_{11} & r_{22} & r_{33} & 0 & \cdots & 0 \\ 0 & r_{12} & r_{23} & \ddots & \ddots & \vdots \\ \vdots & \ddots & r_{13} & \ddots & \ddots & 0 \\ \vdots & & & \ddots & \ddots & \ddots & r_{3n} \\ \vdots & & & & \ddots & \ddots & r_{2n} \\ 0 & \cdots & \cdots & \cdots & 0 & r_{1n} \end{bmatrix}$$

with

$$(2.3.2) \qquad Q_n \hat{J}_n = \begin{bmatrix} R_n \\ 0 \end{bmatrix}.$$

Recall that $\hat{J}_n$ (cf. (2.1.29)) has full rank and thus R_n is nonsingular.

2.3 A Useful Tool

A standard approach for updating the QR factorization (2.3.2) is based on Givens rotations G_n. More precisely, one computes

$$(2.3.3) \qquad Q_n = G_n \cdot \begin{bmatrix} G_{n-1} & 0 \\ 0 & 1 \end{bmatrix} \cdot \begin{bmatrix} G_{n-2} & 0 \\ 0 & I_2 \end{bmatrix} \cdots \begin{bmatrix} G_1 & 0 \\ 0 & I_{n-1} \end{bmatrix}$$

where, for $j = 1, 2, \ldots, n$,

$$(2.3.4) \quad G_j = \begin{bmatrix} I_{j-1} & 0 & 0 \\ 0 & c_j & s_j \\ 0 & -s_j & c_j \end{bmatrix}, \quad \text{with } c_j, s_j \in \mathbb{R} \text{ and } c_j^2 + s_j^2 = 1.$$

Note that

$$(2.3.5) \qquad Q_n = G_n \begin{bmatrix} Q_{n-1} & 0 \\ 0 & 1 \end{bmatrix}, \quad Q_1 = G_1.$$

To devise a formula for the components of R_n and G_n, respectively, we investigate the cases $n = 1, 2, 3$ "by hand". The general case then follows immediately. We have, for $n = 1$,

$$Q_1 \hat{J}_1 = G_1 \hat{J}_1 = \begin{bmatrix} c_1 & s_1 \\ -s_1 & c_1 \end{bmatrix} \begin{bmatrix} \alpha_1 \\ \beta_2 \end{bmatrix} \stackrel{!}{=} \begin{bmatrix} r_{11} \\ 0 \end{bmatrix},$$

and consequently

$$\hat{r}_{11} = \alpha_1, \quad r_{11} = \sqrt{\hat{r}_{11}^2 + \beta_2^2}, \quad c_1 = \frac{\hat{r}_{11}}{r_{11}}, \quad \text{and} \quad s_1 = \frac{\beta_2}{r_{11}}.$$

The case $n = 2$ looks like

$$Q_2 \hat{J}_2 = \begin{bmatrix} 1 & 0 & 0 \\ 0 & c_2 & s_2 \\ 0 & -s_2 & c_2 \end{bmatrix} \begin{bmatrix} r_{11} & c_1 \beta_2 + s_1 \alpha_2 \\ 0 & c_1 \alpha_2 - s_1 \beta_2 \\ 0 & \beta_3 \end{bmatrix} \stackrel{!}{=} \begin{bmatrix} r_{11} & r_{22} \\ 0 & r_{12} \\ 0 & 0 \end{bmatrix}.$$

Hence,

$$r_{22} = c_1 \beta_2 + s_1 \alpha_2, \quad \hat{r}_{12} = c_1 \alpha_2 - s_1 \beta_2, \quad r_{12} = \sqrt{\hat{r}_{12}^2 + \beta_3^2},$$

and

$$c_2 = \frac{\hat{r}_{12}}{r_{12}}, \quad s_2 = \frac{\beta_3}{r_{12}}.$$

Finally, we obtain, for $n = 3$,

$$Q_3 \hat{J}_3 = \begin{bmatrix} 1 & 0 & 0 & 0 \\ 0 & 1 & 0 & 0 \\ 0 & 0 & c_3 & s_3 \\ 0 & 0 & -s_3 & c_3 \end{bmatrix} \begin{bmatrix} r_{11} & r_{22} & s_1\beta_3 \\ 0 & r_{12} & c_1c_2\beta_3 + s_2\alpha_3 \\ 0 & 0 & c_2\alpha_3 - s_2c_1\beta_3 \\ 0 & 0 & \beta_4 \end{bmatrix}$$

$$\overset{!}{=} \begin{bmatrix} r_{11} & r_{22} & r_{33} \\ 0 & r_{12} & r_{23} \\ 0 & 0 & r_{13} \\ 0 & 0 & 0 \end{bmatrix},$$

thus

$$r_{33} = s_1\beta_3, \quad r_{23} = c_1c_2\beta_3 + s_2\alpha_3, \quad \hat{r}_{13} = c_2\alpha_3 - s_2c_1\beta_3,$$

and

$$r_{13} = \sqrt{\hat{r}_{13}^2 + \beta_4^2}, \quad c_3 = \frac{\hat{r}_{13}}{r_{13}}, \quad s_3 = \frac{\beta_4}{r_{13}}.$$

We have our next lemma.

Lemma 2.3.1 *Let $\{\hat{J}_n(\alpha_n, \beta_n)\}$ (cf. (2.1.30)) be given. Then, the upper triangular factor R_n (cf. (2.3.1)) of the QR factorization (2.3.2) and the Givens rotation G_n (cf. (2.3.4)) are given by*

$$r_{1n} = \sqrt{\hat{r}_{1n}^2 + \beta_{n+1}^2}, \quad n \geq 1; \quad r_{3n} = s_{n-2}\beta_n, \quad n \geq 3,$$

$$r_{2n} = \begin{cases} c_1\beta_2 + s_1\alpha_2, & n = 2, \\ c_{n-2}c_{n-1}\beta_n + s_{n-1}\alpha_n, & n \geq 3; \end{cases}$$

$$\hat{r}_{1n} = \begin{cases} \alpha_1, & n = 1, \\ c_1\alpha_2 - s_1\beta_2, & n = 2, \\ c_{n-1}\alpha_n - s_{n-1}c_{n-2}\beta_n, & n \geq 3; \end{cases}$$

$$c_n = \frac{\hat{r}_{1n}}{r_{1n}}, \quad s_n = \frac{\beta_{n+1}}{r_{1n}}, \quad n \geq 1.$$

Recall that the offdiagonal entries β_j of J_n are all positive, as long as an invariant subspace has not been found (cf.(2.1.29)), so that the Givens parameter c_n and s_n are well-defined.

2.4 Orthogonal Residual Polynomials

We will frequently make use of the recursive structure (2.3.5) of Q_n. The next lemma collects some auxiliary results in this direction.

Lemma 2.3.2 Let Q_n and R_n be the orthogonal and upper triangular factors in the QR factorization (2.3.2) of $\hat{J}_n$. Moreover, let $\hat{y} \in \mathbb{R}^{n-1}$ and let $\hat{\eta} \in \mathbb{R}$.

(a) $\quad Q_n^T \begin{bmatrix} \hat{y} \\ \hat{\eta} \\ 0 \end{bmatrix} = \begin{bmatrix} I_n \\ 0 \end{bmatrix} Q_{n-1}^T \begin{bmatrix} \hat{y} \\ 0 \end{bmatrix} + \hat{\eta} Q_n^T \begin{bmatrix} e_n \\ 0 \end{bmatrix}.$

(b) $\quad Q_n^T \begin{bmatrix} e_n \\ 0 \end{bmatrix} = c_n \begin{bmatrix} I_n \\ 0 \end{bmatrix} Q_{n-1}^T e_n + s_n e_{n+1}.$

(c) $\quad Q_n^T \begin{bmatrix} e_n \\ 0 \end{bmatrix} = \hat{J}_n R_n^{-1} e_n.$

(d) $\quad Q_n^T e_{n+1} = -s_n \begin{bmatrix} I_n \\ 0 \end{bmatrix} Q_{n-1}^T e_n + c_n e_{n+1}.$

Proof. The proofs are straightforward applications of (2.3.5) and (2.3.2), respectively. We just indicate that (a)

$$Q_n^T \begin{bmatrix} \hat{y} \\ \hat{\eta} \\ 0 \end{bmatrix} = \begin{bmatrix} Q_{n-1}^T & 0 \\ 0 & 1 \end{bmatrix} \begin{bmatrix} I_{n-1} & 0 & 0 \\ 0 & c_n & -s_n \\ 0 & s_n & c_n \end{bmatrix} \begin{bmatrix} \hat{y} \\ \hat{\eta} \\ 0 \end{bmatrix}$$

$$= \begin{bmatrix} I_n \\ 0 \end{bmatrix} Q_{n-1}^T \begin{bmatrix} \hat{y} \\ 0 \end{bmatrix} + \hat{\eta} \left(c_n \begin{bmatrix} I_n \\ 0 \end{bmatrix} Q_{n-1}^T e_n + s_n e_{n+1} \right)$$

follows from (b). $\square$

Finally, let us introduce the convenient notation

(2.3.6) $\qquad \{Q_n, R_n; \hat{J}_n\} \quad \text{and} \quad \{Q_n, R_n(r_{i,j}), G_n(c_n, s_n); \hat{J}_n\}$

for the QR factorization of $\hat{J}_n$ based on Givens rotations.

2.4 Orthogonal Residual Polynomials

Consider the polynomial based iteration

$$r_n = p_n(A) r_0, \quad p_n(0) = 1.$$

For obvious reasons, one is interested in schemes with short recurrences. Clearly, this is the case if the residual polynomial $p(t) = \psi_n^{\mathrm{OR}}(t)$ is a normalized orthogonal polynomial

$$\psi_n^{\mathrm{OR}}(t) = \frac{\psi_n(t)}{\psi_n(0)}.$$

In this section we investigate such orthogonal residual polynomials. We work out their three-term recurrence relations, discuss the degenerate case $\psi_n(0) = 0$, and characterize ψ_n^{OR} as the solution of a certain least squares problem.

Recurrence Relations for Orthogonal Residual Polynomials

The three-term recurrence relation

$$\gamma_n \psi_n^{\mathrm{OR}}(t) = (t - \alpha_n) \psi_{n-1}^{\mathrm{OR}}(t) - \beta_n \psi_{n-2}^{\mathrm{OR}}(t)$$

for orthogonal residual polynomials $\psi_n^{\mathrm{OR}}(t) = \psi_n(t)/\psi_n(0)$ is special. Here, the interpolatory constraint $\psi_j^{\mathrm{OR}}(0) = 1$ implies

$$\gamma_n = -(\alpha_n + \beta_n).$$

The next lemma characterizes the degenerate case.

Lemma 2.4.1 *Let $\{\psi_j^{\mathrm{OR}}, (\alpha_j, \beta_j, \gamma_j); \langle \cdot, \cdot \rangle\}$ and $\{\psi_j; \langle \cdot, \cdot \rangle\}$ be given. Furthermore, let $\psi_j(0) \neq 0$, $j = 1, 2, \ldots, n-1$, then*

$$\psi_n(0) = 0, \quad \text{if, and only if} \quad \gamma_n = 0.$$

2.4 Orthogonal Residual Polynomials

Proof. The proof makes use of (2.1.22)

$$\gamma_n = -(\alpha_n + \beta_n) = -\frac{\langle t\psi_{n-1}^{OR}, \psi_{n-1}^{OR}\rangle}{\langle \psi_{n-1}^{OR}, \psi_{n-1}^{OR}\rangle} - \frac{\langle t\psi_{n-1}^{OR}, \psi_{n-2}^{OR}\rangle}{\langle \psi_{n-2}^{OR}, \psi_{n-2}^{OR}\rangle}$$

$$= -\langle t\psi_{n-1}, \psi_{n-1}\rangle - \frac{\psi_{n-2}(0)}{\psi_{n-1}(0)}\langle t\psi_{n-1}, \psi_{n-2}\rangle$$

$$= \frac{\langle t\psi_n, \psi_{n-1}\rangle}{\psi_{n-1}(0)}\psi_n(0).$$

Recall that the interlacing property Theorem 2.1.5(b) ensures that ψ_n and ψ_{n-1} can not vanish simultaneously at the origin. □

If $\gamma_n \neq 0$ one may compute the orthogonal residual polynomials by the following the Stieltjes procedure.

Algorithm 2.4.2 Let $\{\psi_j; \langle\cdot,\cdot\rangle\}$ (cf. (2.1.13)) be given. If $\psi_j(0) \neq 0$, $j = 1, 2, \ldots, n$, then the residual polynomials $\{\psi_j^{OR}\}_{j=0}^n$, orthogonal with respect to $\langle\cdot,\cdot\rangle$, may be computed by the following recursive procedure.

Set
$$\psi_{-1}^{OR}(t) := 0, \quad \psi_0^{OR}(t) = 1.$$

For $j = 1, 2, \ldots, n$ compute

$$\alpha_j = \frac{\langle t\psi_{j-1}^{OR}, \psi_{j-1}^{OR}\rangle}{\langle \psi_{j-1}^{OR}, \psi_{j-1}^{OR}\rangle},$$

$$\beta_j = \gamma_{j-1}\frac{\langle \psi_{j-1}^{OR}, \psi_{j-1}^{OR}\rangle}{\langle \psi_{j-2}^{OR}, \psi_{j-2}^{OR}\rangle} \quad (\beta_1 := 0), \quad \gamma_j = -(\alpha_j + \beta_j),$$

and

$$\gamma_j \psi_j^{OR}(t) = (t - \alpha_j)\psi_{j-1}^{OR}(t) - \beta_j \psi_{j-2}^{OR}(t).$$

For a variety of weight functions, such as the *Jacobi weights*, the three-term recurrence coefficients of the corresponding orthogonal polynomials are explicitly known (see, e.g., Szegö [122]). Based on these coefficients, it is not hard to devise a formula for the three-term recurrence coefficients of the associated residual polynomials.

Lemma 2.4.3 *Let the orthogonal polynomials*

$$p_{-1}(t) = 0, \quad p_0(t) = p_0,$$
$$\hat{\gamma}_j p_j(t) = (t - \hat{\alpha}_j)p_{j-1}(t) - \hat{\beta}_j p_{j-2}(t), \quad j \geq 1,$$

be given. Moreover, assume that these polynomials do not vanish at the origin $p_j(0) \neq 0$. Then the three-term recurrence coefficients

$$\gamma_j \psi_j^{\text{OR}}(t) = (t + (\gamma_j + \beta_j))\psi_{j-1}^{\text{OR}}(t) - \beta_j \psi_{j-2}^{\text{OR}}(t).$$

of the associated residual polynomials $\psi_j^{\text{OR}}(t) := p_j(t)/p_j(0)$ are given by

$$\beta_j = \frac{\hat{\beta}_j \hat{\gamma}_{j-1}}{\gamma_{j-1}} \quad (\beta_1 := 0), \quad \text{and} \quad \gamma_j = -(\hat{\alpha}_j + \beta_j).$$

Proof. Rewriting the recurrence relation of the given polynomials yields, for $j \geq 0$,

$$\hat{\gamma}_j \frac{p_j(0)}{p_{j-1}(0)} \frac{p_j(t)}{p_j(0)} = (t - \hat{\alpha}_j)\frac{p_{j-1}(t)}{p_{j-1}(0)} - \hat{\beta}_j \frac{p_{j-2}(0)}{p_{j-1}(0)} \frac{p_{j-2}(t)}{p_{j-2}(0)},$$

or

(2.4.1) $$\gamma_j \psi_j^{\text{OR}}(t) = (t - \hat{\alpha}_j)\psi_{j-1}^{\text{OR}}(t) - \beta_j \psi_{j-2}^{\text{OR}}(t)$$

with

(2.4.2) $$\gamma_j = \hat{\gamma}_j \frac{p_j(0)}{p_{j-1}(0)} \quad \text{and} \quad \beta_j = \hat{\beta}_j \frac{p_{j-2}(0)}{p_{j-1}(0)}.$$

The wanted representation of β_j is obtained by combining the expressions in (2.4.2). Next, by evaluating (2.4.1) at the point $t = 0$ we obtain

$$\gamma_j = -(\hat{\alpha}_j + \beta_j),$$

which concludes the proof. □

2.4 Orthogonal Residual Polynomials

Extremal Property of Orthogonal Residual Polynomials

The orthogonal residual polynomials ψ_n^{OR} are uniquely characterized by a certain optimality property. It is this condition which makes the CG method, when applied to a positive definite system, such a powerful scheme (compare Example 2.4.8).

Before we state this result, let us further investigate the degenerate case $\psi_n(0) = 0$.

Lemma 2.4.4 *Let $\{\psi_n, \nu_0, J_n; \langle \cdot, \cdot \rangle\}$ (cf. (2.1.30)) be given. Then, the following conditions are equivalent:*

(a) $\psi_n(0) = 0$;

(b) J_n is singular;

(c) $J_n y = \sqrt{\nu_0} e_1$ has no solution.

Proof. The equivalence of (a) and (b) follows from Theorem 2.1.6. We now show that the linear system in (c) has no solution if the coefficient matrix is singular. To see this, let us investigate the case $n = 4$, i.e., let J_4 be singular. First, recall from Theorem 2.1.5(b) that J_3 is regular. Now, assume to the contrary that the linear system has a solution, which is only possible, if it is consistent

$$3 = \operatorname{rank} J_3 = \operatorname{rank} J_4 = \operatorname{rank} \begin{bmatrix} \alpha_1 & \beta_2 & 0 & 0 & \sqrt{\nu_0} \\ \beta_2 & \alpha_2 & \beta_3 & 0 & 0 \\ 0 & \beta_3 & \alpha_3 & \beta_4 & 0 \\ 0 & 0 & \beta_4 & \alpha_4 & 0 \end{bmatrix}.$$

These rank requirements imply that the vector $\sqrt{\nu_0} e_1$ is in the range of the first 3 columns of J_4. This, however, is impossible, since all $\beta_k > 0$, $k = 1, 2, 3, 4$ (cf. (2.1.23)). □

One may ask, why are we interested in the system stated in Lemma 2.4.4(c). Part (c) of the next theorem gives an answer to this question.

Theorem 2.4.5 Let $\{\psi_n, \nu_0, J_n; \langle \cdot, \cdot \rangle\}$ (cf. (2.1.30)) be given. If $\psi_n(0) \neq 0$, then the following conditions are equivalent:

(a) $\quad \psi_n^{\mathrm{OR}}(t) = \dfrac{\psi_n(t)}{\psi_n(0)};$

(b) $\quad \langle \psi_n^{\mathrm{OR}}, q \rangle = 0$ for all $q \in \Pi_{n-1}, \quad \psi_n^{\mathrm{OR}}(0) = 1;$

(c) $\quad \psi_n^{\mathrm{OR}}(t) = 1 - t \Psi_{n-1}^{\mathrm{T}}(t) y_n^{\mathrm{OR}},$

where y_n^{OR} is the solution of $J_n y = \sqrt{\nu_0} e_1$.

Proof. Let $q \in \Pi_{n-1}$. We apply (2.1.28) to obtain, for $x, y \in \mathbb{R}^n$,

$$q(t) = \Psi_{n-1}^{\mathrm{T}}(t) x \quad \text{and} \quad \psi_n^{\mathrm{OR}}(t) = 1 - \Psi_n^{\mathrm{T}}(t) \hat{J}_n y.$$

From this and the orthogonality of the ψ_j's we deduce

$$\langle \psi_n^{\mathrm{OR}}, q \rangle = \langle 1 - \Psi_n^{\mathrm{T}}(t) \hat{J}_n y, \Psi_{n-1}^{\mathrm{T}}(t) x \rangle$$
$$= x^{\mathrm{T}} \left(\sqrt{\nu_0} e_1 - J_n y \right)$$

and the statements follow. $\square$

For future reference we summarize the various expressions of the residual polynomial ψ_n^{OR}, orthogonal with respect to $\langle \cdot, \cdot \rangle$, as obtained from Theorem 2.1.5 and Theorem 2.4.5

(2.4.3) $\quad \psi_n^{\mathrm{OR}}(t) = \dfrac{\psi_n(t)}{\psi_n(0)} = 1 - t \Psi_{n-1}^{\mathrm{T}}(t) y_n^{\mathrm{OR}} = \prod_{j=1}^{n} \left(1 - \dfrac{t}{\theta_j^{(n)}} \right).$

It turns out that the solution y_n^{OR} of the linear system (c) is the stationary point of a quadratic form. Indeed, let $p(t) = 1 - t \Psi_{n-1}^{\mathrm{T}}(t) y$ be some residual polynomial and define

(2.4.4) $\quad f(y) := \langle p, t^{-1} p \rangle = \langle 1 - t \Psi_{n-1}^{\mathrm{T}}(t) y, t^{-1} - \Psi_{n-1}^{\mathrm{T}}(t) y \rangle$
$$= \langle 1, t^{-1} \rangle + y^{\mathrm{T}} J_n y - 2 y^{\mathrm{T}} \sqrt{\nu_0} e_1.$$

2.4 Orthogonal Residual Polynomials

Then, it holds
$$f'(y_n^{OR}) = 2(J_n y_n^{OR} - \sqrt{\nu_0} e_1) = 0,$$

with

(2.4.5) $$f(y_n^{OR}) = \langle 1, t^{-1} \rangle - \nu_0 e_1^T J_n^{-1} e_1.$$

Here, we implicitly assumed that the "modified inner product" $\langle \cdot, \cdot t^{-1} \rangle$ is well-defined, which is certainly the case if the origin does not belong to the supporting set of $\langle \cdot, \cdot \rangle$.

Formula (2.4.5) is a familiar expression. With the help of Theorem 2.2.2 we obtain as a first corollary to Theorem 2.4.5:

Corollary 2.4.6 *Let $\{\psi_n, \nu_0, J_n; \langle \cdot, \cdot \rangle\}$ (cf. (2.1.30)) be given. If the associated orthogonal residual polynomial ψ_n^{OR} exists, then the error in the Gaussian quadrature rule applied to the function t^{-1} reads*
$$\langle \psi_n^{OR}, t^{-1} \psi_n^{OR} \rangle = \langle 1, t^{-1} \rangle - \nu_0 e_1^T J_n^{-1} e_1.$$

By discussing the Hessian J_n of the quadratic form (2.4.5) we are in a position to state the optimality properties of ψ_n^{OR}.

Corollary 2.4.7 *Let $\{J_n; \langle \cdot, \cdot \rangle\}$ (cf. (2.1.30)) be given. Moreover, let $\psi_n^{OR}(t) = 1 - t\Psi_{n-1}^T(t) y_n^{OR}$ -if existing - denote the associated orthogonal residual polynomial.*

(a) If J_n is positive definite, then
$$\langle \psi_n^{OR}, t^{-1} \psi_n^{OR} \rangle = \min \{ \langle p, t^{-1} p \rangle : p \in \Pi_n, \, p(0) = 1 \}.$$

(b) If J_n is indefinite, then y_n^{OR} is the unique saddlepoint of the quadratic form defined in (2.4.4).

(c) If J_n is singular, then ψ_n^{OR} does not exist.

We conclude this section by showing that the CG method is a particular application of the theory of orthogonal residual polynomials.

Example 2.4.8 Let ψ_n^{OR} denote the orthogonal residual polynomial with respect to the inner product (cf. (1.1.11))
$$\langle p, q \rangle_{\text{GAL}} = r_0^T p(A) q(A) r_0.$$
In this example we discuss properties of the associated polynomial iteration method

(2.4.6) $\qquad r_n^{\text{OR}} = \psi_n^{\text{OR}}(A) r_0, \quad \text{or} \quad \varepsilon_n^{\text{OR}} = \psi_n^{\text{OR}}(A) \varepsilon_0.$

As already stated in the introduction, the orthogonality of the ψ_n^{OR}'s yields the well-known *Galerkin condition* for the residuals

(2.4.7) $\quad 0 = \langle \psi_k^{\text{OR}}, \psi_l^{\text{OR}} \rangle_{\text{GAL}} = r_0^T \psi_k^{\text{OR}}(A) \psi_l^{\text{OR}}(A) r_0 = (r_k^{\text{OR}})^T r_l^{\text{OR}}, \quad k \neq l.$

Moreover, the quadratic form (2.4.4) is nothing but the square of the A-norm of the error

(2.4.8)
$$\begin{aligned} \langle \psi_n^{\text{OR}}, t^{-1} \psi_n^{\text{OR}} \rangle_{\text{GAL}} &= r_0^T \psi_n^{\text{OR}}(A) A^{-1} \psi_n^{\text{OR}}(A) r_0 \\ &= \varepsilon_0^T \psi_n^{\text{OR}}(A) A \psi_n^{\text{OR}}(A) \varepsilon_0 \\ &= (\varepsilon_n^{\text{OR}})^T A \varepsilon_n^{\text{OR}} \\ &= \|\varepsilon_n^{\text{OR}}\|_A^2 \end{aligned}$$

Hence, Cor. 2.4.7(a) states that the method defined by (2.4.6) minimizes the A-norm of the error
$$\|\varepsilon_n^{\text{OR}}\|_A = \min \left\{ \|p(A)\varepsilon_0\|_A : p \in \Pi_n, \, p(0) = 1 \right\}.$$
for positive definite A. In other words, the CG method of Hestenes and Stiefel [81] is defined by the orthogonal residual polynomials ψ_n^{OR} (for an implementation see Section 6.2).

If A symmetric and indefinite, the situation is more complicated. Here, the Galerkin iterate x_n^{OR} need not to exist for every n (cf. Lemma 2.4.4) and the quadratic form, in general, (2.4.8) no longer has minimum (cf. Cor. 2.4.7). In fact, the "A-norm" is no longer a norm.

Finally, we learn from Cor. 2.4.6, that the square of the A-norm of the error may be interpreted as the error in the Gaussian quadrature rule when applied to the initial approximation

(2.4.9) $\qquad \|\varepsilon_n^{\text{OR}}\|_A^2 = \|\varepsilon_0^{\text{OR}}\|_A^2 - \nu_0 e_1^T J_n^{-1} e_1,$

a fact, already known to Stiefel [115, p.164]. $\qquad \square$

2.5 Kernel Polynomials

We just learned from Example 2.4.8 that the CG method minimizes the A-norm of the error, a quantity which is usually not at hand. It seems to be more natural to minimize instead the Euclidian norm of the residual

$$\|r_n\|_2 = \min\{\|p(A)r_0\|_2 : p \in \Pi_n, \, p(0) = 1\}.$$

Again, it is possible to explicitly determine the solution of this least squares problem. This time, the solution is given by a particular class of orthogonal residual polynomials, the so-called kernel polynomials. Their properties will be discussed in the next subsections.

Definition and Orthogonality

Again, we denote by $\{\psi_j\}_{j=0}^n$ the system of orthonormal polynomials with respect to $\langle \cdot, \cdot \rangle$. Let $\xi \in \mathbb{R}$ be a fixed number and define the polynomials

$$(2.5.1) \qquad K_n(t;\xi) := \sum_{j=0}^{n} \psi_j(t)\psi_j(\xi) = \Psi_n^T(t)\Psi_n(\xi).$$

K_n is called the *kernel polynomial* with respect to $\langle \cdot, \cdot \rangle$ (and the parameter ξ). Note that

$$(2.5.2) \qquad K_n(t;\xi) \text{ has exact degree } \begin{cases} n & \text{if } \psi_n(\xi) \neq 0, \\ n-1 & \text{if } \psi_n(\xi) = 0. \end{cases}$$

The name "kernel" is motivated by the following result, which is also known as the *reproducing property* of the kernel polynomials

$$(2.5.3) \quad \langle K_n(t;\xi), p(t) \rangle = \int_a^d K_n(t;\xi)p(t)d\sigma(t) = p(\xi), \quad \text{for all} \quad p \in \Pi_n.$$

To prove (2.5.3) one simply applies the Fourier expansion (2.1.18) of p. It is worth noticing that $K_n(t;\xi)$ is uniquely determined by the reproducing property (see, e.g., Davis [28, §10.1]). In particular, the property (2.5.3) holds true for all polynomials $p(t) = (t - \xi)q(t)$ with zero ξ. Thus, we have as an important result

$$(2.5.4) \qquad \langle K_n(t;\xi), q(t)(t - \xi) \rangle = 0 \quad \text{for all} \quad q \in \Pi_{n-1}.$$

In other words, $K_n(t;\xi)$ is orthogonal with respect to the "modified" inner product $\langle \cdot, \cdot \cdot (t - \xi) \rangle$.

Extremal Property of Kernel Polynomials

In this section we investigate the kernel polynomials $K_n(t;\xi)/K_n(\xi;\xi)$. Since $\psi_0(t) > 0$ the denominator $K_n(\xi;\xi) = \sum_{j=0}^{n} \psi_j^2(\xi) > 0$ is always positive (cf. (2.5.1)). The scaled kernel polynomials $K_n(t;\xi)/K_n(\xi;\xi)$ have a significant extremal property which is important for the design of efficient polynomial acceleration methods. These schemes always involve the *constraint point* $\xi = 0$ (cf. (1.1.2)). Therefore, we assume, without loss of generality, that $\xi = 0$ is fixed. Clearly, the general case $\xi \in \mathbb{R}$ can be reduced to this special case by means of the linear shift $t - \xi$.

The proof for the fact, that the scaled kernel polynomials solve a certain approximation problem can be found for example in Chihara [22, Theorem 7.3]. We offer a different proof which provides also a useful formula for the computation of kernel polynomials.

Theorem 2.5.1 *Let $\{\psi_n, \nu_0, \hat{J}_n; \langle \cdot, \cdot \rangle\}$ (cf. (2.1.30)) and the associated kernel polynomial $K_n(t;0)$ (cf. (2.5.1)) be given. Then, the following conditions are equivalent:*

(a) $\quad \psi_n^{\mathrm{MR}}(t) = \dfrac{K_n(t;0)}{K_n(0;0)}$;

(b) $\quad \langle \psi_n^{\mathrm{MR}}, tp \rangle = 0$ *for all* $p \in \Pi_{n-1}$, $\psi_n^{\mathrm{MR}}(0) = 1$;

(c) $\quad \|\psi_n^{\mathrm{MR}}\| = \min\{\|p\| : p \in \Pi_n,\ p(0) = 1\}$, $\psi_n^{\mathrm{MR}}(0) = 1$;

(d) $\quad \psi_n^{\mathrm{MR}}(t) = 1 - t\Psi_{n-1}^{\mathrm{T}}(t) y_n^{\mathrm{MR}}$,

where y_n^{MR} is the solution of $\hat{J}_n^{\mathrm{T}} \hat{J}_n y = \sqrt{\nu_0} \hat{J}_n^{\mathrm{T}} e_1$.

Proof. That (a) is equivalent to (b) follows directly from (2.5.4). Let $q \in \Pi_{n-1}$. From (2.1.28) we obtain

$$tq(t) = \Psi_n^{\mathrm{T}}(t)\hat{J}_n x \quad \text{and} \quad \psi_n^{\mathrm{MR}}(t) = 1 - \Psi_n^{\mathrm{T}}(t)\hat{J}_n y,$$

where $x, y \in \mathbb{R}^n$. From this

$$\langle \psi_n^{\mathrm{MR}}, tq \rangle = \langle 1 - \Psi_n^{\mathrm{T}}(t)\hat{J}_n y, \Psi_n^{\mathrm{T}}(t)\hat{J}_n x \rangle$$
$$= x^{\mathrm{T}}\left(\sqrt{\nu_0}\hat{J}_n^{\mathrm{T}} e_1 - \hat{J}_n^{\mathrm{T}} \hat{J}_n y\right),$$

2.5 Kernel Polynomials

which clearly shows that (b) is equivalent to (d). To show that (c) is equivalent to (d) observe that

$$f(y) := \|\psi_n^{\text{MR}}\|^2 = \langle 1 - \Psi_n^T(t)\hat{J}_n y, 1 - \Psi_n^T(t)\hat{J}_n y\rangle$$
$$= \langle 1, 1\rangle + y^T \hat{J}_n^T \hat{J}_n y - 2\sqrt{\nu_0} y^T \hat{J}_n^T e_1.$$

Thus, the zero y_n^{MR} of the gradient $f'(y_n^{\text{MR}}) = 0$ is just the solution of the normal equations (d). This and the fact that the Hessian $\hat{J}_n^T \hat{J}_n$ is positive definite (cf. (2.1.29)) concludes the proof. □

We remark that the superscript MR in the expressions

(2.5.5) $$\psi_n^{\text{MR}}(t) = \frac{K_n(t;0)}{K_n(0;0)} = 1 - t\Psi_{n-1}^T(t)y_n^{\text{MR}}.$$

indicates that ψ_n^{MR} will define the minimal residual method as is apparent from the next example.

Example 2.5.2 Let ψ_n^{MR} denote the scaled kernel polynomial with respect to the inner product (compare Example 2.4.8)

$$\langle p, q\rangle_{\text{GAL}} = r_0^T p(A) q(A) r_0.$$

Let us investigate the resulting polynomial iteration method

$$r_n^{\text{MR}} = \psi_n^{\text{MR}}(A) r_0.$$

Observe that the orthogonality property of the ψ_n^{MR}'s translates into the so-called *Petrov-Galerkin condition* for the residuals

$$0 = \langle \psi_k^{\text{MR}}, t\psi_l^{\text{MR}}\rangle_{\text{GAL}} = r_0^T \psi_k^{\text{MR}}(A) A \psi_l^{\text{MR}}(A) r_0 = (r_k^{\text{MR}})^T A r_l^{\text{MR}}.$$

So, we recovered the *Conjugate Residual method* (CR) due to Stiefel [115, Satz 4] (see, also Rutishauser [106]). In fact, Stiefel [116] seems to be the first who applied the theory of kernel polynomials to the iterative solution of linear systems.

Next, we point out that the CR method minimizes the Euclidian norm of the residual (cf. Theorem 2.5.1(c))

$$\|r_n^{\text{MR}}\|_2 = \min\{\|p(A) r_0\|_2 : p \in \Pi_n, \ p(0) = 1\},$$

i.e., ψ_n^{MR} defines a polynomial iteration method with **Minimal Residual**.

Finally, it is interesting to note that the degenerate case $\psi_n(0) = 0$ implies that $\psi_n^{\text{MR}} = \psi_{n-1}^{\text{MR}}$, i.e., the iteration process stagnates $\|r_n^{\text{MR}}\|_2 = \|r_{n-1}^{\text{MR}}\|_2$. $\square$

It is in general not advisable to solve directly the normal equations in Theorem 2.5.1(d). This problem can be avoided by means of the QR decomposition (2.3.2).

Corollary 2.5.3 *Let $\{\psi_n, \nu_0, \hat{J}_n; \langle \cdot, \cdot \rangle\}$ (cf. (2.1.30)) and the QR factorization $\{Q_n, R_n, G_n(c_n, s_n); \hat{J}_n\}$ (cf. (2.3.6)) be given. Furthermore, let $\psi_n^{\text{MR}}(t) = 1 - t\Psi_{n-1}^T(t)y_n^{\text{MR}}$ denote the scaled kernel polynomial (2.5.5).*

(a) The solution y_n^{MR} of the normal equations in Theorem 2.5.1(d) is given by

$$R_n y_n^{\text{MR}} = \hat{y}_n^{\text{MR}}, \quad \text{where} \quad \begin{bmatrix} \hat{y}_n^{\text{MR}} \\ \hat{\eta}_n^{\text{MR}} \end{bmatrix} := \sqrt{\nu_0} Q_n e_1.$$

Moreover, the following update formulae hold

$$\hat{y}_n^{\text{MR}} = \begin{bmatrix} \hat{y}_{n-1}^{\text{MR}} \\ c_n \hat{\eta}_{n-1}^{\text{MR}} \end{bmatrix} \quad \text{and} \quad \hat{\eta}_n^{\text{MR}} = -s_n \hat{\eta}_{n-1}^{\text{MR}}, \quad \hat{\eta}_1^{\text{MR}} = -\sqrt{\nu_0} s_1.$$

(b) The norm of ψ_n^{MR} is given by

$$\|\psi_n^{\text{MR}}\| = \sqrt{\nu_0} \prod_{j=1}^{n} |s_j| = |s_n| \|\psi_{n-1}^{\text{MR}}\|.$$

Proof. For (a) we start with the normal equations and apply the QR decomposition (2.3.2)

2.5 Kernel Polynomials

$$0 = \hat{J}_n^T \left(\hat{J}_n y_n^{MR} - \sqrt{\nu_0} e_1 \right)$$
$$= (Q_n \hat{J}_n)^T \left(Q_n \hat{J}_n y_n^{MR} - \sqrt{\nu_0} Q_n e_1 \right)$$
$$= [\, R_n^T \;\; 0\,] \left(\begin{bmatrix} R_n \\ 0 \end{bmatrix} y_n^{MR} - \begin{bmatrix} \hat{y}_n^{MR} \\ \hat{\eta}_n^{MR} \end{bmatrix} \right)$$
$$= R_n^T (R_n y_n^{MR} - \hat{y}_n^{MR}).$$

Furthermore, we apply (2.3.5) and obtain

$$\begin{bmatrix} \hat{y}_n^{MR} \\ \hat{\eta}_n^{MR} \end{bmatrix} = \sqrt{\nu_0} Q_n e_1 = G_n \begin{bmatrix} \sqrt{\nu_0} Q_{n-1} e_1 \\ 0 \end{bmatrix} = G_n \begin{bmatrix} \hat{y}_{n-1}^{MR} \\ \hat{\eta}_{n-1}^{MR} \\ 0 \end{bmatrix} = \begin{bmatrix} \hat{y}_{n-1}^{MR} \\ c_n \hat{\eta}_{n-1}^{MR} \\ -s_n \hat{\eta}_{n-1}^{MR} \end{bmatrix}.$$

To compute the norm we make use of (2.1.28) and of part (a)

$$\psi_n^{MR}(t) = 1 - \Psi_n^T(t) \hat{J}_n y_n^{MR}$$
$$= \Psi_n^T(t) \left(\sqrt{\nu_0} e_1 - \hat{J}_n y_n^{MR} \right)$$
$$= \Psi_n^T(t) Q_n^T \left(\sqrt{\nu_0} Q_n e_1 - Q_n \hat{J}_n y_n^{MR} \right)$$
$$= \hat{\eta}_n^{MR} \Psi_n^T(t) Q_n^T e_{n+1}.$$

Consequently, we derive

$$\|\psi_n^{MR}\|^2 = (\hat{\eta}_n^{MR})^2 \langle \Psi_n^T(t) Q_n^T e_{n+1}, \Psi_n^T(t) Q_n^T e_{n+1} \rangle$$
$$= (\hat{\eta}_n^{MR})^2 e_{n+1}^T Q_n Q_n^T e_{n+1}$$
$$= (\hat{\eta}_n^{MR})^2,$$

which concludes the proof. □

We remark that, as a by-product, the proof provides an alternative expression for the residual polynomial

(2.5.6) $$\psi_n^{MR}(t) = \hat{\eta}_n^{MR} \Psi_n^T(t) Q_n^T e_{n+1}.$$

Recurrence Relations for Kernel Polynomials

In this subsection we work out recurrence relations for the residual polynomials ψ_n^{MR}. We will investigate three different versions.

The first one is the classical three-term recurrence relation

$$(2.5.7) \qquad \gamma_j \psi_j^{MR}(t) = (t + (\gamma_j + \beta_j))\psi_{j-1}^{MR}(t) - \beta_j \psi_{j-2}^{MR}(t),$$

based on the fact (cf. (2.5.4)) that the ψ_n^{MR} are orthogonal with respect to the modified weight function $\langle \cdot, \cdot \, t \rangle$. This approach is fine as long as the origin does not belong to the interval of orthogonality $0 \notin (a,d)$. Otherwise, the corresponding weight function $tw(t)$ (or the discrete analog) changes sign in $[a, d]$. Such weight functions were studied by Struble [120]. The trouble is that orthogonal polynomials with respect to *variable-signed weight functions* need not to exist for any degree. Hence, a Stieltjes-like algorithm for the computation of the recurrence coefficients in (2.5.7) may break down. Incidentally, the degenerate case is again characterized by $\psi_n(0) = 0$.

To devise computable recurrence relations which work as well in the degenerate case we will discuss two "mixed recurrence relations"

$$(2.5.8) \qquad \psi_n^{MR}(t) = \rho_n \psi_{n-1}^{MR}(t) - \rho_{n-1} t p_{n-1}(t),$$

and

$$(2.5.9) \qquad \psi_n^{MR}(t) = \hat{\rho}_{n-1} \psi_{n-1}^{MR}(t) - \hat{\rho}_n p_n(t).$$

Here p_n and p_{n-1} are suitable chosen polynomials of exact degree n and $n-1$, respectively, and ρ_{n-1}, ρ_n, $\hat{\rho}_{n-1}$, $\hat{\rho}_n \in \mathbb{R}$ are parameter to be determined.

It is worth noticing that each representation of ψ_n^{MR} leads to a different, though mathematically equivalent, implementation of the conjugate residual method described in Example 2.5.2. The first representation (2.5.7) produces the classical CR implementation (see, e.g., Reid [104]). The first mixed relation (2.5.8) defines the MCR (**M**odified **C**onjugate **R**esidual) algorithm of Chandra [18], [19]. Finally, relation (2.5.9) leads to the MINRES (**MIN**imal **RES**idual) implementation of Paige and Saunders [95].

Our first result characterizes the degenerate case.

2.5 Kernel Polynomials

Lemma 2.5.4 Let $\{\psi_n, ; \langle \cdot, \cdot \rangle\}$ (cf. (2.1.13)) and $\{G_n(c_n, s_n); \hat{J}_n\}$ (cf. (2.3.6)) be given. Furthermore, let $\psi_{n-1}^{MR}(t)$ denote the scaled kernel polynomial (2.5.5).

(a) There exists an orthogonal polynomial of exact degree n with respect to $\langle \cdot, \cdot \, t \rangle$ if, and only if $\psi_n(0) \neq 0$.

(b) $c_n \neq 0$ if, and only if $\psi_n(0) \neq 0$.

(c) If $\psi_n(0) = 0$ then $\langle \psi_{n-1}^{MR}, tp \rangle = 0$ for all $p \in \Pi_{n-1}$.

Proof. (a) If $\psi_n(0) \neq 0$, then it follows from (2.5.2) and Theorem 2.5.1(b) that ψ_n^{MR} has degree n and satisfies the required orthogonality relation.

Now, let $\psi_n(0) = 0$ and assume to the contrary that p_n has exact degree n with
$$\langle p_n, tq \rangle = 0, \quad \text{for all} \quad q \in \Pi_{n-1}.$$
Then there exist constants ρ_n and $\rho_{n+1} \neq 0$ with

(2.5.10) $$tp_n(t) = \rho_{n+1}\psi_{n+1}(t) + \rho_n\psi_n(t).$$

The assumption immediately implies $\psi_{n+1}(0) = 0$, which is a contradiction to Theorem 2.1.5(b).

The proof of (b) and (c) relies on the fact that $\psi_n(0) = 0$ is equivalent to $\psi_n^{MR}(t) \equiv \psi_{n-1}^{MR}(t)$ (cf. (2.5.1)). In the light of this observation, part (b) follows from $s_n^2 + c_n^2 = 1$ (cf. 2.3.4) and Cor. 2.5.3(b), whereas part (c) is a consequence of Theorem 2.5.1(b)
$$0 = \langle \psi_n^{MR}, tp \rangle = \langle \psi_{n-1}^{MR}, tp \rangle,$$
which concludes the proof. □

Notice that statement (c) is in particular valid for $p \equiv \psi_{n-1}^{MR}$. This rules out the application of (2.1.17) for the generation of recursion coefficients for ψ_n^{MR}. On the other hand, however, if $\psi_j(0) \neq 0$, $j = 1, 2, \ldots, n$, then part (a) ensures that ψ_n^{MR} may be computed by a Stieltjes-like algorithm.

> **Algorithm 2.5.5** Let $\{\psi_j; \langle \cdot, \cdot \rangle\}$ (cf. (2.1.13)) be given. If $\psi_j(0) \neq 0$, $j = 1, 2, \ldots, n$, then the residual polynomials $\{\psi_j^{\mathrm{MR}}\}_{j=0}^n$, orthogonal with respect to $\langle \cdot, t \cdot \rangle$, may be computed by the following recursive procedure.
>
> Set
> $$\psi_{-1}^{\mathrm{MR}}(t) := 0, \quad \psi_0^{\mathrm{MR}}(t) = 1.$$
>
> For $j = 1, 2, \ldots, n$ compute
> $$\alpha_j = \frac{\langle t\psi_{j-1}^{\mathrm{MR}}, t\psi_{j-1}^{\mathrm{MR}} \rangle}{\langle \psi_{j-1}^{\mathrm{MR}}, t\psi_{j-1}^{\mathrm{MR}} \rangle},$$
>
> $$\beta_j = \gamma_{j-1} \frac{\langle \psi_{j-1}^{\mathrm{MR}}, t\psi_{j-1}^{\mathrm{MR}} \rangle}{\langle \psi_{j-2}^{\mathrm{MR}}, t\psi_{j-2}^{\mathrm{MR}} \rangle} \quad (\beta_1 := 0), \quad \gamma_j = -(\alpha_j + \beta_j),$$
>
> and
> $$\gamma_j \psi_j^{\mathrm{MR}}(t) = (t - \alpha_j)\psi_{j-1}^{\mathrm{MR}}(t) - \beta_j \psi_{j-2}^{\mathrm{MR}}(t).$$

Let us now turn to (2.5.8), i.e., let us determine a polynomial p_{n-1} and numbers ρ_n, ρ_{n-1} with

$$\psi_n^{\mathrm{MR}}(t) = \rho_n \psi_{n-1}^{\mathrm{MR}}(t) - \rho_{n-1} t p_{n-1}(t)$$

Obviously, the interpolatory constraint $\psi_n^{\mathrm{MR}}(0) = 1$ determines $\rho_n = 1$. It remains to adjust p_{n-1} and ρ_{n-1} such that

(2.5.11) $\qquad 0 = \langle \psi_n^{\mathrm{MR}}, tp_j \rangle = \langle \psi_{n-1}^{\mathrm{MR}}, tp_j \rangle - \rho_{n-1} \langle p_{n-1}, t^2 p_j \rangle,$

for $j = 1, 2, \ldots, n-1$, where the p_j's are polynomials of exact degree j. For $j < n-1$, the residual polynomial ψ_n^{MR} satisfies (2.5.11), for any ρ_{n-1}, if p_{n-1} is orthogonal with respect to the modified weight function $\langle \cdot, \cdot t^2 \rangle$. The case $j = n-1$ uniquely determines the remaining parameter

$$\rho_{n-1} = \frac{\langle \psi_{n-1}^{\mathrm{MR}}, tp_{n-1} \rangle}{\langle p_{n-1}, t^2 p_{n-1} \rangle}.$$

2.5 Kernel Polynomials

Note that this time the inner product $\langle \cdot, \cdot t^2 \rangle$ does not change sign and hence the denominator of ρ_{n-1} never vanishes. Moreover, in view of Lemma 2.5.4(c), the degenerate case $\psi_n(0) = 0$ just implies $\rho_{n-1} = 0$.

In principle p_{n-1} could be any orthogonal polynomial with respect to $\langle \cdot, \cdot t^2 \rangle$. In our algorithms we always have chosen $p_{n-1}(t) \equiv \psi_{n-1}^{\text{MO}}(t)$ to be the associated monic orthogonal polynomial.

So, the next algorithm generates beside the residual polynomial ψ_n^{MR} as well the monic polynomials ψ_n^{MO}, orthogonal with respect to $\langle \cdot, \cdot t^2 \rangle$.

Algorithm 2.5.6 *The residual polynomials* $\{\psi_j^{\text{MR}}\}_{j=0}^n$, *orthogonal with respect to* $\langle \cdot, \cdot t \rangle$ *and the monic polynomials* $\{\psi_j^{\text{MO}}\}_{j=0}^{n-1}$, *orthogonal with respect to* $\langle \cdot, \cdot t^2 \rangle$, *may be computed by the following recursive procedure.*

Set
$$\psi_0^{\text{MR}}(t) = 1, \quad \psi_{-1}^{\text{MO}}(t) := 0, \quad \psi_0^{\text{MO}}(t) = 1.$$

For $j = 1, 2, \ldots, n$ compute

$$\rho_{j-1} = \frac{\langle \psi_{j-1}^{\text{MR}}, t\psi_{j-1}^{\text{MO}} \rangle}{\langle \psi_{j-1}^{\text{MO}}, t^2 \psi_{j-1}^{\text{MO}} \rangle}, \quad \psi_j^{\text{MR}}(t) = \psi_{j-1}^{\text{MR}}(t) - \rho_{j-1} t \psi_{j-1}^{\text{MO}}(t),$$

$$\alpha_j = \frac{\langle t\psi_{j-1}^{\text{MO}}, t^2 \psi_{j-1}^{\text{MO}} \rangle}{\langle \psi_{j-1}^{\text{MO}}, t^2 \psi_{j-1}^{\text{MO}} \rangle}, \quad \beta_j = \frac{\langle \psi_{j-1}^{\text{MO}}, t^2 \psi_{j-1}^{\text{MO}} \rangle}{\langle \psi_{j-2}^{\text{MO}}, t^2 \psi_{j-2}^{\text{MO}} \rangle}, \quad \beta_1 := 0,$$

$$\psi_j^{\text{MO}}(t) = (t - \alpha_j)\psi_{j-1}^{\text{MO}}(t) - \beta_j \psi_{j-2}^{\text{MO}}(t).$$

Finally, let us discuss the recurrence relation stated in (2.5.9). This time the recurrence coefficients are computed via the QR factorization (2.3.2). A similar derivation can be found in Freund [51, Theorem 4.1]. Part (b) of the following theorem will turn out to be useful for devising update formulae for the associated polynomial acceleration method. Also, it constitutes another version of (2.5.8).

Theorem 2.5.7 Let $\{\psi_n, \hat{J}_n; \langle \cdot, \cdot \rangle\}$ (cf. (2.1.30)) and the QR factorization $\{Q_n, R_n(r_{i,j}), G_n(c_n, s_n); \hat{J}_n\}$ (cf. (2.3.6)) be given. Furthermore, let ψ_n^{MR} denote the residual polynomial (2.5.5) and let $\hat{\eta}_n^{\mathrm{MR}}$ be as defined in Cor. 2.5.3. Then the following recurrence relations hold.

(a)
$$\psi_n^{\mathrm{MR}}(t) = s_n^2 \psi_{n-1}^{\mathrm{MR}}(t) + c_n \hat{\eta}_n^{\mathrm{MR}} \psi_n(t).$$

(b)
$$\psi_n^{\mathrm{MR}}(t) = \psi_{n-1}^{\mathrm{MR}}(t) - c_n \hat{\eta}_{n-1}^{\mathrm{MR}} t w_n^{\mathrm{MR}}(t),$$

where
$$w_n^{\mathrm{MR}}(t) := \Psi_{n-1}(t) R_n^{-1} e_n.$$

Moreover, the following update formulae hold

$$w_1^{\mathrm{MR}}(t) = \psi_0(t)/r_{11},$$
$$w_2^{\mathrm{MR}}(t) = (\psi_1(t) - r_{22} w_1^{\mathrm{MR}}(t))/r_{12},$$
$$w_n^{\mathrm{MR}}(t) = (\psi_{n-1}(t) - r_{3n} w_{n-2}^{\mathrm{MR}}(t) - r_{2n} w_{n-1}^{\mathrm{MR}}(t))/r_{1n}, \quad n \geq 3.$$

Proof. Using (2.5.5), (2.1.28) and Cor. 2.5.3(a) we obtain that

$$(2.5.12) \quad \psi_n^{\mathrm{MR}}(t) = 1 - \Psi_n^{\mathrm{T}}(t) Q_n \begin{bmatrix} \hat{y}_n^{\mathrm{MR}} \\ 0 \end{bmatrix} = 1 - \Psi_n^{\mathrm{T}}(t) Q_n^{\mathrm{T}} \begin{bmatrix} \hat{y}_{n-1}^{\mathrm{MR}} \\ c_n \hat{\eta}_{n-1}^{\mathrm{MR}} \\ 0 \end{bmatrix}.$$

This together with Lemma 2.3.2(a) yields

$$(2.5.13) \quad \psi_n^{\mathrm{MR}}(t) = 1 - \Psi_{n-1}^{\mathrm{T}}(t) Q_{n-1}^{\mathrm{T}} \begin{bmatrix} \hat{y}_{n-1}^{\mathrm{MR}} \\ 0 \end{bmatrix} - c_n \hat{\eta}_{n-1}^{\mathrm{MR}} \Psi_n^{\mathrm{T}}(t) Q_n^{\mathrm{T}} \begin{bmatrix} e_n \\ 0 \end{bmatrix}$$
$$= \psi_{n-1}^{\mathrm{MR}}(t) - c_n \hat{\eta}_{n-1}^{\mathrm{MR}} \Psi_n^{\mathrm{T}}(t) Q_n^{\mathrm{T}} \begin{bmatrix} e_n \\ 0 \end{bmatrix}.$$

Finally, we apply Lemma 2.3.2(b), Cor. 2.5.3(a) and (2.5.6) to obtain

$$\psi_n^{\mathrm{MR}}(t) = \psi_{n-1}^{\mathrm{MR}}(t) - c_n \hat{\eta}_{n-1}^{\mathrm{MR}} (c_n \Psi_{n-1}^{\mathrm{T}}(t) Q_{n-1}^{\mathrm{T}} e_n + s_n \psi_n(t))$$
$$= \psi_{n-1}^{\mathrm{MR}}(t) - c_n^2 \psi_{n-1}^{\mathrm{MR}}(t) - c_n s_n \hat{\eta}_{n-1}^{\mathrm{MR}} \psi_n(t)$$
$$= s_n^2 \psi_{n-1}^{\mathrm{MR}}(t) + c_n \hat{\eta}_n^{\mathrm{MR}} \psi_n(t).$$

2.5 Kernel Polynomials

Alternatively, we may apply Lemma 2.3.2(c) and (2.1.28) onto (2.5.13) which then proves the first part of (b). The update formulae for w_n^{MR} follow from comparing related columns of

$$[\psi_0(t) \quad \psi_1(t) \quad \cdots \quad \psi_{n-1}(t)] = [w_1^{MR}(t) \quad w_2^{MR}(t) \quad \cdots \quad w_n^{MR}(t)] R_n.$$

□

Zeros of Kernel Polynomials

In this section we prove a separation theorem for the zeros of kernel polynomials. In particular we compare these zeros with the zeros of the underlying orthogonal polynomials ψ_n.

An alternative representation of the kernel polynomials turns out to be convenient for the discussion of the zeros. The next expression is known as *Christoffel - Darboux Identity* and follows readily from the three-term recurrence representation (2.1.15) (compare Szegö [122, Theorem 3.2.2]), or could be developed by further investigating (2.5.10) (compare Gautschi [62, Theorem 2.1])

$$(2.5.14) \qquad K_n(t;\xi) = \beta_{n+2} \frac{\psi_{n+1}(t)\psi_n(\xi) - \psi_n(t)\psi_{n+1}(\xi)}{t-\xi}.$$

For future reference, we state the special case $t = \xi$

$$(2.5.15) \quad 0 < \sum_{j=0}^{n} \psi_j^2(t) = K_n(t;t) = \beta_{n+2}(\psi'_{n+1}(t)\psi_n(t) - \psi'_n(t)\psi_{n+1}(t)).$$

In the light of representation (2.5.14) it should come as no surprise that the zeros of $K_n(t;\xi)$ are closely related to the zeros of ψ_n and ψ_{n+1}, respectively.

We remark that the next theorem can be found as well in a recent paper by Paige, Parlett and van der Vorst [93, §7]. They, however, used a different method of proof. Also, the special case $\xi \notin [a,d]$ was treated by Chihara [22, Theorem 7.2].

Theorem 2.5.8 Let $\{\psi_n, K_n(t;\xi); \langle \cdot, \cdot \rangle\}$ (cf. (2.1.13)) be given. Furthermore, let $\{\theta_j^{(n)}\}_{j=1}^n$ and $\{\theta_j^{(n+1)}\}_{j=1}^{n+1}$ denote the zeros of ψ_n and ψ_{n+1}, respectively. Finally, define $\theta_0^{(n)} := -\infty$ and $\theta_{n+1}^{(n)} := \infty$.

(a) If $\xi = \theta_k^{(n)}$ is a zero of ψ_n, then $K_n(t;\xi)$ has the $n-1$ zeros

$$\theta_1^{(n)} < \theta_2^{(n)} < \cdots < \theta_{k-1}^{(n)} < \theta_{k+1}^{(n)} < \theta_{k+2}^{(n)} < \cdots < \theta_n^{(n)}.$$

(b) If $\xi = \theta_k^{(n+1)}$ is a zero of ψ_{n+1}, then $K_n(t;\xi)$ has the n zeros

$$\theta_1^{(n+1)} < \theta_2^{(n+1)} < \cdots < \theta_{k-1}^{(n+1)} < \theta_{k+1}^{(n+1)} < \theta_{k+2}^{(n+1)} < \cdots < \theta_{n+1}^{(n+1)}.$$

(c) Let $\xi \in (\theta_k^{(n+1)}, \theta_k^{(n)})$, $k \in \{1, 2, \ldots, n+1\}$, then $K_n(t;\xi)$ has precisely one zero in each interval

$$(\theta_j^{(n+1)}, \theta_j^{(n)}), \quad j = 1, 2, \ldots, k-1, k+1, k+2, \ldots, n+1.$$

(d) Let $\xi \in (\theta_{k-1}^{(n)}, \theta_k^{(n+1)})$, $k \in \{1, 2, \ldots, n+1\}$, then $K_n(t;\xi)$ has precisely one zero in each interval

$$(\theta_{j-1}^{(n)}, \theta_j^{(n+1)}), \quad j = 1, 2, \ldots, k-1, k+1, k+2, \ldots, n+1.$$

Proof. Part (a) and (b) are an immediate consequence of (2.5.14). For the proof of part (c) and (d), respectively, we have to collect some auxiliary results. First, recall from Theorem 2.1.5(b) that the zeros of ψ_n and ψ_{n+1} separate each other

$$\theta_0^{(n)} < \theta_1^{(n+1)} < \theta_1^{(n)} < \theta_2^{(n+1)} \cdots < \theta_n^{(n)} < \theta_{n+1}^{(n+1)} < \theta_{n+1}^{(n)}.$$

Since ψ_n and ψ_{n+1} have positive leading coefficients, it follows that

(2.5.16)
$$\operatorname{sgn}\left(\psi_{n+1}(\theta_j^{(n)})\right) = (-1)^{j+n+1}, \quad j = 0, 1, \ldots, n+1,$$
$$\operatorname{sgn}\left(\psi_n(\theta_j^{(n+1)})\right) = (-1)^{j+n+1}, \quad j = 1, 2, \ldots, n+1,$$

2.5 Kernel Polynomials

and consequently
(2.5.17)
$$\operatorname{sgn}\left(\frac{\psi_{n+1}(\xi)}{\psi_n(\xi)}\right) = \begin{cases} +1 & \text{if } \xi \in (\theta_k^{(n+1)}, \theta_k^{(n)}), \\ -1 & \text{if } \xi \in (\theta_{k-1}^{(n)}, \theta_k^{(n+1)}), \end{cases} \quad k = 1, 2, \ldots, n+1.$$

Obviously,
$$Q_n(t;\xi) := \frac{1}{\beta_{n+2}\psi_n(\xi)} K_n(t;\xi)$$

and $K_n(t;\xi)$ have the same zeros. (2.5.16) and (2.5.17) determine the sign of
$$P_n(t;\xi) := (t-\xi)Q_n(t;\xi) = \psi_{n+1}(t) - \frac{\psi_{n+1}(\xi)}{\psi_n(\xi)}\psi_n(t).$$

We have

(2.5.18)
$$\operatorname{sgn}\left(P_n(\theta_j^{(n)};\xi)\right) = (-1)^{j+n+1},$$
$$\operatorname{sgn}\left(P_n(\theta_j^{(n+1)};\xi)\right) = \begin{cases} (-1)^{j+n} & \text{if } \xi \in (\theta_k^{(n+1)}, \theta_k^{(n)}) \\ (-1)^{j+n+1} & \text{if } \xi \in (\theta_{k-1}^{(n)}, \theta_k^{(n+1)}) \end{cases}.$$

We are now in a position to prove part (c) and (d). We start with (c), i.e., we assume that $\xi \in (\theta_k^{(n+1)}, \theta_k^{(n)})$. It follows from (2.5.18) that

$$\operatorname{sgn}\left(P_n(\theta_j^{(n+1)};\xi)\right) \neq \operatorname{sgn}\left(P_n(\theta_j^{(n)};\xi)\right), \quad j = 1, 2, \ldots, n+1.$$

Clearly, the polynomial $Q_n(t;\xi) = P_n(t;\xi)/(t-\xi)$ shares this sign property, but for $j = k$

$$\operatorname{sgn}\left(Q_n(\theta_k^{(n+1)};\xi)\right) = \operatorname{sgn}\left(Q_n(\theta_k^{(n)};\xi)\right),$$

which concludes the proof of (c). For (d), i.e., $\xi \in (\theta_{k-1}^{(n)}, \theta_k^{(n+1)})$, we have

$$\operatorname{sgn}\left(P_n(\theta_{j-1}^{(n)};\xi)\right) \neq \operatorname{sgn}\left(P_n(\theta_j^{(n+1)};\xi)\right), \quad j = 1, 2, \ldots, n+1$$

and
$$\operatorname{sgn}\left(Q_n(\theta_{k-1}^{(n)};\xi)\right) = \operatorname{sgn}\left(Q_n(\theta_k^{(n+1)};\xi)\right),$$

and the wanted sign distribution follows. □

We stress two important parts of the preceding theorem. If $\xi \in (\theta_k^{(n+1)}, \theta_k^{(n)})$, $k \in \{1, 2, \ldots, n\}$, then there is *no* zero of $K_n(t; \xi)$ in

$$[\theta_{k-1}^{(n)}, \theta_{k+1}^{(n+1)}] = [\theta_{k-1}^{(n)}, \theta_k^{(n+1)}] \cup [\theta_k^{(n+1)}, \theta_k^{(n)}] \cup [\theta_k^{(n)}, \theta_{k+1}^{(n+1)}].$$

Moreover, there is precisely one zero of $K_n(t; \xi)$ in $(\theta_{n+1}^{(n+1)}, \infty)$ and no zero in $(-\infty, \theta_1^{(n+1)})$.

For $\xi \in (\theta_{k-1}^{(n)}, \theta_k^{(n+1)})$, $k \in \{2, 3, \ldots, n+1\}$, a similar result holds. This time, $K_n(t; \xi)$ has *no* zero in

$$[\theta_{k-1}^{(n+1)}, \theta_k^{(n)}] = [\theta_{k-1}^{(n+1)}, \theta_{k-1}^{(n)}] \cup [\theta_{k-1}^{(n)}, \theta_k^{(n+1)}] \cup [\theta_k^{(n+1)}, \theta_k^{(n)}],$$

precisely one zero in $(-\infty, \theta_1^{(n+1)})$ and no zero in $(\theta_{n+1}^{(n+1)}, \infty)$.

A little example might be helpful for the verification of the statements in Theorem 2.5.8.

	k	$\theta_0^{(3)}$	$\theta_1^{(4)}$	$\theta_1^{(3)}$	$\theta_2^{(4)}$	$\theta_2^{(3)}$	$\theta_3^{(4)}$	$\theta_3^{(3)}$	$\theta_4^{(4)}$	$\theta_4^{(3)}$
sgn(P_3)		+	+	−	−	+	+	−	−	+
sgn(Q_3)	1	−	−	−	−	+	+	−	−	+
sgn(Q_3)	2	−	−	+	+	+	+	−	−	+
sgn(Q_3)	3	−	−	+	+	−	−	−	−	+
sgn(Q_3)	4	−	−	+	+	−	−	+	+	+

Table 2.5.9 Sign of P_3 and Q_3 at the zeros of ψ_3 and ψ_4 for $\xi \in (\theta_k^{(4)}, \theta_k^{(3)})$

	k	$\theta_0^{(3)}$	$\theta_1^{(4)}$	$\theta_1^{(3)}$	$\theta_2^{(4)}$	$\theta_2^{(3)}$	$\theta_3^{(4)}$	$\theta_3^{(3)}$	$\theta_4^{(4)}$	$\theta_4^{(3)}$
sgn(P_3)		+	−	−	+	+	−	−	+	+
sgn(Q_3)	1	−	−	−	+	+	−	−	+	+
sgn(Q_3)	2	−	+	+	+	+	−	−	+	+
sgn(Q_3)	3	−	+	+	−	−	−	−	+	+
sgn(Q_3)	4	−	+	+	−	−	+	+	+	+

Table 2.5.10 Sign of P_3 and Q_3 at the zeros of ψ_3 and ψ_4 for $\xi \in (\theta_{k-1}^{(3)}, \theta_k^{(4)})$

We finish with the remark that Szegö seems to be the first who was interested in the location of the zeros of kernel functions (see, Szegö [121, §5]). He, however, treated the more general case of complex valued weight functions.

2.5 Kernel Polynomials

Computing Zeros of Kernel Polynomials

The zeros of orthogonal polynomials can be conveniently computed as eigenvalues of the associated Jacobi matrix. The zeros of kernel polynomials turn out to be the eigenvalues of a generalized eigenvalue problem.

The next theorem is due to Freund [51, Corollary 5.3] (compare also, Manteuffel and Otto [89, Theorem 2.3]).

Theorem 2.5.11 Let $\{K_n(t;0), J_n, \hat{J}_n; \langle \cdot, \cdot \rangle\}$ (cf. (2.1.30)) be given. Then, any zero $\hat{\theta}_j^{(n)}$ of $K_n(t;0)$ is an eigenvalue of the generalized eigenvalue problem
$$\hat{J}_n^T \hat{J}_n \hat{s} = \hat{\theta}_j^{(n)} J_n \hat{s}.$$

Proof. Let $\hat{\theta}$ be a zero of $K_n(t;0)$, i.e.,
$$K_n(t;0) = (\hat{\theta} - t)\Psi_{n-1}^T(t)\hat{s}, \quad \hat{s} \in \mathbb{R}^n.$$
The orthogonality of the kernel polynomials (cf. (2.5.4)) and (2.1.28) imply
$$\begin{aligned}
0 &= \langle (\hat{\theta} - t)\Psi_{n-1}^T(t)\hat{s}, t\Psi_{n-1}^T(t)y \rangle \\
&= \hat{\theta} \langle \Psi_{n-1}^T(t)\hat{s}, \Psi_n^T(t)\hat{J}_n y \rangle - \langle \Psi_n^T(t)\hat{J}_n \hat{s}, \Psi_n^T(t)\hat{J}_n y \rangle \\
&= y^T(\hat{\theta} J_n \hat{s} - \hat{J}_n^T \hat{J}_n \hat{s}),
\end{aligned}$$
for all $y \in \mathbb{R}^n$. $\square$

Following the lines of Freund [51], it is again possible to avoid the use of the possibly ill-conditioned matrix $\hat{J}_n^T \hat{J}_n$. The trick is to multiply this matrix by the inverse of the upper triangular factor of the QR factorization (2.3.2)
$$R_n^{-T} \hat{J}_n^T \hat{J}_n = R_n^{-T} \hat{J}_n^T Q_n^T Q_n \hat{J}_n = R_n^{-T} [R_n^T \ 0] \begin{bmatrix} R_n \\ 0 \end{bmatrix} = R_n.$$

Likewise, we obtain for the right-hand side of the generalized eigenvalue problem
$$R_n^{-T} J_n = R_n^{-T} \hat{J}_n^T \begin{bmatrix} I_n \\ 0 \end{bmatrix} = [I_n \ 0] Q_n \begin{bmatrix} I_n \\ 0 \end{bmatrix} =: \hat{Q}_n.$$

By exploiting the recursive structure of Q_n (cf. (2.3.3)) the following corollary results.

Corollary 2.5.12 *Let $\{K_n(t;0), \hat{J}_n; \langle \cdot, \cdot \rangle\}$ (cf. (2.1.30)) and the QR factorization $\{Q_n, R_n, G_n(c_n, s_n); \hat{J}_n\}$ (cf. (2.3.6)) be given. Then, any zero $\hat{\theta}_j^{(n)}$ of $K_n(t;0)$ is an eigenvalue of the generalized eigenvalue problem*

$$R_n \hat{s} = \hat{\theta}_j^{(n)} \hat{Q}_n \hat{s},$$

where

$$\hat{Q}_n = \begin{bmatrix} I_{n-1} & 0 \\ 0 & c_n \end{bmatrix} \cdot G_{n-1} \cdot \begin{bmatrix} G_{n-2} & 0 \\ 0 & 1 \end{bmatrix} \cdots \begin{bmatrix} G_1 & 0 \\ 0 & I_{n-2} \end{bmatrix}, \quad \hat{Q}_1 = c_1.$$

2.6 Hermite Kernel Polynomials

In the previous sections we have investigated the residual polynomials ψ_n^{OR} and ψ_n^{MR}, respectively. Both polynomials need not exist if the origin belongs to the interval of orthogonality. In this section we will define a new set of residual polynomials which overcomes this shortcoming. We obtain these polynomials by relaxing the orthogonality condition. To uniquely define them, we then add one more constraint, namely we require these polynomials to have a zero derivative at the origin. On account of this interpolatory constraint we will denote them as *Hermite kernel polynomials*.

To our best knowledge, the Hermite kernel polynomials have not yet been investigated in the literature.

In view of the fact that the scaled kernel polynomials minimize the Euclidian norm of the residual, it is tempting to look for polynomials which minimize the Euclidian norm of the error

$$\|\varepsilon_n\|_2 = \min \{\|p(A)\varepsilon_0\|_2 : p \in \Pi_n, \, p(0) = 1\}.$$

Unfortunately, the solution of this problem requires the knowledge of x_*, the solution of the given linear system (compare Section 6.7). In this situation, Fridman [53] suggested to start the associated polynomial iteration method

2.6 Hermite Kernel Polynomials

with Ar_0 instead of r_0. Now, the approximation problem above changes to (compare Example 2.6.2)

$$\|\varepsilon_n\|_2 = \min\{\|p(A)\varepsilon_0\|_2 : p \in \Pi_{n+1},\ p(0) = 1,\ p'(0) = 0\}.$$

Here, the solution is provided by the Hermite kernel polynomials.

There is yet another motivation for the introduction of the additional constraint $p'(0) = 0$. It arises in the design of polynomial based iteration methods for the solution of *singular and inconsistent linear systems* (compare Fischer, Hanke and Hochbruck [43] and Calvetti, Reichel and Zhang [15]). To see this, let us assume for a moment that A is singular and inconsistent. Here, the righthand side

$$f = f_\mathcal{R} + f_\mathcal{N}$$

splits into components in the range of A and in the nullspace of A, respectively. Furthermore, let $x_* = A^\dagger f$ denote the unique solution satisfying

$$Ax_* = f_\mathcal{R} \quad \text{and} \quad x_* - x_0 \in \text{range}(A).$$

Moreover, recall from (1.1.3) that

$$x_n = x_0 + q_{n-1}(A)r_0$$

for $r_n = p_n(A)r_0$ with $p_n(t) = 1 - tq_{n-1}(t)$. Working from here a straightforward calculation yields

$$\begin{aligned}\varepsilon_n &= x_* - x_n \\ &= x_* - p_n(A)x_0 - q_{n-1}(A)f \\ &= p_n(A)(x_* - x_0) - q_{n-1}(A)f_\mathcal{N} \\ &= p_n(A)\varepsilon_0 - q_{n-1}(0)f_\mathcal{N}.\end{aligned}$$

Hence, the condition $p'_n(0) = 0$, or equivalently the condition $q_{n-1}(0) = 0$, ensures that the iterates x_n converge to x_* for appropriate residual polynomials p_n.

Definition and Orthogonality

Let us first define an auxiliary set of polynomials which may be viewed as "kernel polynomials of kernel polynomials"

$$(2.6.1) \qquad \hat{K}_n(t;\xi) := \sum_{j=0}^n \psi_j(t)\psi'_j(\xi),$$

where $\{\psi_j\}_{j=0}^n$ denotes the system of orthonormal polynomials with respect to $\langle \cdot, \cdot \rangle$ and $\xi \in \mathbb{R}$ is a fixed parameter. Note that (cf. (2.5.1))

$$\hat{K}_n(\xi;\xi) = K'_n(\xi;\xi). \tag{2.6.2}$$

The degree of $\hat{K}_n(t;\xi)$, $n \geq 1$, depends on the location of ξ

$$\hat{K}_n(t;\xi) \text{ has exact degree } \begin{cases} n & \text{if } \psi'_n(\xi) \neq 0, \\ n-1 & \text{if } \psi'_n(\xi) = 0. \end{cases} \tag{2.6.3}$$

Here we used the fact that, in view of (2.5.15), the derivatives of ψ'_n and ψ'_{n-1} do not vanish at the same point. Moreover, Theorem 2.1.5(a) implies that the degenerate case $\psi'_n(\xi) = 0$ may only occur for $\xi \in (a, d)$.

The polynomials defined in (2.6.1) satisfy a reproducing property (compare (2.5.3)). We have, for all $p \in \Pi_{n-1}$,

$$\langle \hat{K}_n(t;\xi), (t-\xi)p(t) \rangle = \int_a^d \hat{K}_n(t;\xi)p(t)(t-\xi)d\sigma(t) = p(\xi), \tag{2.6.4}$$

To prove (2.6.4) we apply the Fourier expansion (2.1.18)

$$(t-\xi)p(t) = \sum_{j=0}^n \langle \psi_j(t), (t-\xi)p(t) \rangle \psi_j(t)$$

and deduce

$$\langle \hat{K}_n(t;\xi), (t-\xi)p(t) \rangle = \sum_{j=0}^n \langle \psi_j(t), (t-\xi)p(t) \rangle \psi'_j(\xi)$$
$$= [(t-\xi)p(t)]'_{t=\xi}$$
$$= p(\xi).$$

In particular, formula (2.6.4) holds true for all polynomials $p(t) = (t-\xi)q(t)$ with zero ξ. Thus, as a by-product, the modified kernel polynomials fulfill an orthogonality relation

$$\langle \hat{K}_n(t;\xi), q(t)(t-\xi)^2 \rangle = 0 \quad \text{for all} \quad q \in \Pi_{n-2}.$$

Next, combining (2.6.4) with (2.5.3) yields

$$\langle (t-\xi)\hat{K}_n(t;\xi) - K_n(t;\xi), p(t) \rangle = 0, \quad \text{for all} \quad p \in \Pi_{n-1}.$$

2.6 Hermite Kernel Polynomials

In other words, the polynomial

(2.6.5) $$K_n^{\text{H}}(t;\xi) := (t-\xi)\hat{K}_n(t;\xi) - K_n(t;\xi)$$

is orthogonal over Π_{n-1}. We stress that $K_n^{\text{H}} \in \Pi_{n+1}$ is in general not orthogonal with respect to polynomials of degree n. K_n^{H} is called *Hermite kernel polynomial* with respect to $\langle\cdot,\cdot\rangle$. The name is partly motivated by the fact that, as an immediate consequence of (2.6.2), the derivative of K_n^{H} vanishes at the parameter ξ

$$(K_n^{\text{H}})'(\xi;\xi) = \left[(t-\xi)\hat{K}_n'(t;\xi) + \hat{K}_n(t;\xi) - K_n'(t;\xi)\right]_{t=\xi} = 0.$$

Also, it is worth noticing that K_n^{H} itself does not vanish at the parameter ξ

(2.6.6) $$K_n^{\text{H}}(\xi;\xi) = -K_n(\xi;\xi) \neq 0.$$

Extremal Property and a Christoffel-Darboux Formula

We have already learned that scaled orthogonal polynomials ψ_n^{OR} and scaled kernel polynomials ψ_n^{MR} are characterized by an extremal property. One may ask whether this is also true for scaled Hermite kernel polynomials $K_n^{\text{H}}(t;\xi)/K_n^{\text{H}}(\xi;\xi)$. Notice that (2.6.6) implies that the denominator $K_n^{\text{H}}(\xi;\xi)$ does not vanish. Again, the most important case is defined by the special parameter $\xi = 0$. The next theorem shows that the scaled Hermite kernel polynomials indeed solve an approximation problem.

Before we state the result, let us rewrite our "standard inner product" $\langle p,q\rangle_{\text{GAL}} = r_0^{\text{T}} p(A) q(A) r_0$ in terms of the error

(2.6.7) $$\langle t^{-1}p, t^{-1}q\rangle_{\text{GAL}} = \varepsilon_0^{\text{T}} p(A)q(A)\varepsilon_0,$$

which is possible as long as A is not singular. Clearly, the new inner product defines a norm

(2.6.8) $$\|t^{-1}p\|_{\text{GAL}} := \sqrt{\langle t^{-1}p, t^{-1}p\rangle_{\text{GAL}}} = \|p(A)\varepsilon_0\|_2.$$

Theorem 2.6.1 Let $\{\psi_n, \nu_0, \hat{J}_n; \langle \cdot, \cdot \rangle\}$ (cf. (2.1.30)) and the associated Hermite kernel polynomial $K_n^H(t;0)$ (cf. (2.6.5)) be given. Then, the following conditions are equivalent:

(a) $\psi_n^{ME}(t) = \dfrac{K_n^H(t;0)}{K_n^H(0;0)}$;

(b) $\langle \psi_n^{ME}, p \rangle = 0$ for all $p \in \Pi_{n-1}$, $\psi_n^{ME}(0) = 1$, $(\psi_n^{ME})'(0) = 0$;

(c) $\|t^{-1}\psi_n^{ME}\| = \min\{\|t^{-1}p\| : p \in \Pi_{n+1},\ p(0) = 1,\ p'(0) = 0\}$,
    with $\psi_n^{ME}(0) = 1$, $(\psi_n^{ME})'(0) = 0$;

(d) $\psi_n^{ME}(t) = 1 - t^2 \Psi_{n-1}^T(t) y_n^{ME}$,

where y_n^{ME} is the solution of $\hat{J}_n^T \hat{J}_n y = \sqrt{\nu_0} e_1$.

Proof. The proof is completely analogous to the corresponding one for kernel polynomials. Only the part (b) $\Rightarrow$ (a), i.e., ψ_n^{ME} is uniquely determined by the interpolatory constraints and by (b), needs some further investigation. If ψ_n^{ME} fulfills (b), then there exist real numbers ρ_{n+1} and ρ_n with

$$\psi_n^{ME}(t) = \rho_{n+1}\psi_{n+1}(t) + \rho_n \psi_n(t).$$

The constraints on ψ_n^{ME} lead to the linear system

(2.6.9) $$\begin{bmatrix} \psi_{n+1}(0) & \psi_n(0) \\ \psi_{n+1}'(0) & \psi_n'(0) \end{bmatrix} \begin{bmatrix} \rho_{n+1} \\ \rho_n \end{bmatrix} = \begin{bmatrix} 1 \\ 0 \end{bmatrix},$$

which has a unique solution, since the determinant

$$\psi_{n+1}(0)\psi_n'(0) - \psi_{n+1}'(0)\psi_n(0) \neq 0$$

does not vanish as follows from (2.5.15). □

Here, the superscript ME in the expressions

(2.6.10) $$\psi_n^{ME}(t) := \dfrac{K_n^H(t;0)}{K_n^H(0;0)} = 1 - t^2 \Psi_{n-1}^T(t) y_n^{ME}.$$

2.6 Hermite Kernel Polynomials

indicates that ψ_n^{ME} will define a Minmal Error method, as is apparent from the next example.

Example 2.6.2 Let ψ_n^{ME} denote the scaled Hermite kernel polynomial with respect to the inner product (compare Example 2.4.8 and Example 2.5.2)

$$\langle p, q \rangle_{\mathrm{GAL}} = r_0^{\mathrm{T}} p(A) q(A) r_0.$$

This polynomial gives rise to the polynomial acceleration method

$$r_n^{\mathrm{ME}} = \psi_n^{\mathrm{ME}}(A) r_0 \quad \text{or} \quad \varepsilon_n^{\mathrm{ME}} = \psi_n^{\mathrm{ME}}(A) \varepsilon_0$$

which, in view of (2.6.7), (2.6.8) and Theorem 2.6.1(c), minimizes the Euclidian norm of the error

$$\|\varepsilon_n^{\mathrm{ME}}\|_2 = \min \left\{ \|p(A)\varepsilon_0\|_2 : p \in \Pi_{n+1},\ p(0) = 1,\ p'(0) = 0 \right\}.$$

This time, the orthogonality property of the ψ_n^{ME}'s translates to

$$0 = \langle \psi_n^{\mathrm{ME}}, p \rangle_{\mathrm{GAL}} = r_n^{\mathrm{ME}} p(A) r_0 = \varepsilon_n^{\mathrm{ME}} A p(A) \varepsilon_0, \quad \text{for all} \quad p \in \Pi_{n-1},$$

or, equivalently,

$$r_n^{\mathrm{ME}} \perp \mathcal{K}_n(A; r_0) \quad \text{and} \quad \varepsilon_n^{\mathrm{ME}} \perp \mathcal{K}_n(A; A r_0).$$

Again, the degenerate case $\psi_n'(0) = 0$ results in a stagnation of the iterative process

$$\|r_n^{\mathrm{ME}}\|_2 = \|r_{n-1}^{\mathrm{ME}}\|_2$$

(compare Example 2.5.2). □

In accordance with the "MR-case" (cf. Cor. 2.5.3) it is possible to avoid the direct solve of the normal equations developed in Theorem 2.6.1(d), and it is possible to provide an update formula for the norm of ψ_n^{ME}.

Corollary 2.6.3 Let $\{\psi_n, \nu_0, \hat{J}_n; \langle \cdot, \cdot \rangle\}$ (cf. (2.1.30)) and the QR factorization $\{Q_n, R_n(r_{i,j}), G_n(c_n, s_n); \hat{J}_n\}$ (cf. (2.3.6)) be given. Furthermore, let $\psi_n^{\mathrm{ME}}(t) = 1 - t^2 \Psi_{n-1}^{\mathrm{T}}(t) y_n^{\mathrm{ME}}$ denote the scaled Hermite kernel polynomial (2.6.10).

(a) The solution y_n^{ME} of the normal equations in Theorem 2.6.1(d) is given by

$$R_n y_n^{\mathrm{ME}} = \hat{y}_n^{\mathrm{ME}}, \quad \text{where} \quad R_n^{\mathrm{T}} \hat{y}_n^{\mathrm{ME}} = \sqrt{\nu_0} e_1.$$

Moreover, the following update formulae hold

$$\hat{y}_n^{\mathrm{ME}} = \begin{bmatrix} \hat{y}_{n-1}^{\mathrm{ME}} \\ \hat{\eta}_n^{\mathrm{ME}} \end{bmatrix}, \quad \hat{y}_1^{\mathrm{ME}} = \hat{\eta}_1^{\mathrm{ME}},$$

where

$$\hat{\eta}_1^{\mathrm{ME}} = \sqrt{\nu_0}/r_{11},$$
$$\hat{\eta}_2^{\mathrm{ME}} = -r_{22}\hat{\eta}_1^{\mathrm{ME}}/r_{12},$$
$$\hat{\eta}_n^{\mathrm{ME}} = -(r_{3n}\hat{\eta}_{n-2}^{\mathrm{ME}} + r_{2n}\hat{\eta}_{n-1}^{\mathrm{ME}})/r_{1n}, \quad n \geq 3.$$

(b) The norm of ψ_n^{ME} is given by

$$\|\psi_n^{\mathrm{ME}}\| = \sqrt{(r_{1,n+1}\hat{\eta}_{n+1}^{\mathrm{ME}})^2 + (r_{3,n+2}\hat{\eta}_n^{\mathrm{ME}})^2}.$$

Proof. To derive (a) we apply the QR factorization (2.3.2) onto the normal equations Theorem 2.6.1(d)

$$\sqrt{\nu_0} e_1 = \hat{J}_n^T \hat{J}_n y_n^{\mathrm{ME}} = (Q_n \hat{J}_n)^{\mathrm{T}} (Q_n \hat{J}_n) y_n^{\mathrm{ME}} = R_n^{\mathrm{T}} R_n y_n^{\mathrm{ME}}.$$

We "motivate" the update formula for $\hat{y}_n^{\mathrm{ME}}$ by a little example

2.6 Hermite Kernel Polynomials

$$R_3^T \hat{y}_3^{ME} = \begin{bmatrix} r_{11} & 0 & 0 \\ r_{22} & r_{12} & 0 \\ r_{33} & r_{23} & r_{13} \end{bmatrix} \begin{bmatrix} \hat{y}_2^{ME} \\ \hat{\eta}_3^{ME} \end{bmatrix}$$

$$= \begin{bmatrix} R_2^T \hat{y}_2^{ME} \\ r_{33}\hat{\eta}_1^{ME} + r_{23}\hat{\eta}_2^{ME} + r_{13}\hat{\eta}_3^{ME} \end{bmatrix} = \sqrt{\nu_0} \begin{bmatrix} e_1 \\ 0 \end{bmatrix}.$$

(b) The formula is based upon the fact that ψ_n^{ME} can be expressed as a linear combination of the orthonormal polynomials ψ_n and ψ_{n+1}. We apply (a), (2.1.28), (2.3.2), the fact $\psi_0(t) = \nu_0^{-1/2}$, and obtain

$$\psi_n^{ME}(t) = 1 - t^2 \Psi_{n-1}^T(t) y_n^{ME}$$
$$= 1 - t^2 \Psi_{n-1}^T(t) R_n^{-1} \hat{y}_n^{ME}$$
$$= 1 - \Psi_{n+1}^T(t) \hat{J}_{n+1} \hat{J}_n R_n^{-1} \hat{y}_n^{ME}$$
$$= 1 - \Psi_{n+1}^T(t) \hat{J}_{n+2}^T \hat{J}_{n+2} \begin{bmatrix} I_n \\ 0 \\ 0 \end{bmatrix} R_n^{-1} \hat{y}_n^{ME}$$
$$= 1 - \Psi_{n+1}^T(t) R_{n+2}^T R_{n+2} \begin{bmatrix} R_n^{-1} \hat{y}_n^{ME} \\ 0 \\ 0 \end{bmatrix}$$
$$= 1 - \sqrt{\nu_0}\Psi_{n+1}^T(t) e_1 + \Psi_{n+1}^T(t) R_{n+2}^T R_{n+2} \left(y_{n+2}^{ME} - \begin{bmatrix} R_n^{-1} \hat{y}_n^{ME} \\ 0 \\ 0 \end{bmatrix} \right)$$
$$= \Psi_{n+1}^T(t) R_{n+2}^T \begin{bmatrix} 0 \\ \hat{\eta}_{n+1}^{ME} \\ \hat{\eta}_{n+2}^{ME} \end{bmatrix}$$
$$= \Psi_{n+1}^T(t) \begin{bmatrix} 0 \\ r_{1,n+1}\hat{\eta}_{n+1}^{ME} \\ r_{2,n+2}\hat{\eta}_{n+1}^{ME} + r_{1,n+2}\hat{\eta}_{n+2}^{ME} \end{bmatrix}$$
$$= (r_{1,n+1}\hat{\eta}_{n+1}^{ME})\psi_n(t) + (r_{3,n+2}\hat{\eta}_n^{ME})\psi_{n+1}(t),$$

which concludes the proof. □

As an interesting side result we notice that the auxiliary polynomial $\hat{K}_n$ (cf. (2.6.1)) essentially describes the difference between ψ_n^{ME} and ψ_n^{MR}. Indeed, from (2.5.5), (2.6.5), and (2.6.10), we get

(2.6.11) $$\psi_n^{\text{ME}}(t) = \psi_n^{\text{MR}}(t) - t\frac{\hat{K}_n(t;0)}{K_n(0;0)}.$$

To discuss the degenerate case $\psi_n'(0) = 0$, i.e., the case where ψ_n^{ME} has not full degree (cf. (2.6.3)), it is cumbersome to investigate one of the representations of ψ_n^{ME} written down in (2.6.10). Instead, we explicitly solve the linear system defined in (2.6.9) and obtain the following expression for the residual polynomials

(2.6.12) $$\psi_n^{\text{ME}}(t) = \frac{\psi_n'(0)\psi_{n+1}(t) - \psi_{n+1}'(0)\psi_n(t)}{\psi_n'(0)\psi_{n+1}(0) - \psi_{n+1}'(0)\psi_n(0)}.$$

By considering this representation for n and for $n-1$, respectively, one easily verifies the following result.

Corollary 2.6.4 *Let $\{\psi_n; \langle \cdot, \cdot \rangle\}$ (cf. (2.1.13)), with associated residual polynomial $\psi_n^{\text{OR}}(t) = \psi_n(t)/\psi_n(0)$, be given. Then*

$$\psi_n'(0) = 0 \quad \text{if, and only if} \quad \psi_n^{\text{ME}}(t) \equiv \psi_n^{\text{OR}}(t).$$

The representation (2.6.12) may be viewed as a "Christoffel-Darboux formula" for scaled Hermite kernel polynomials (cf. (2.5.14)). Such a formula is interesting on its own right. Therefore we state the following, more general, theorem.

Theorem 2.6.5 *Let $\{\psi_n, (\alpha_n, \beta_n); \langle \cdot, \cdot \rangle\}$ (cf. (2.1.13)) be given. Then the Hermite kernel polynomial may be written as*

$$K_n^{\text{H}}(t;\xi) = \beta_{n+2}(\psi_n'(\xi)\psi_{n+1}(t) - \psi_{n+1}'(\xi)\psi_n(t)).$$

2.6 Hermite Kernel Polynomials

Proof. The proof is straightforward. We just indicate the main step which makes use of the three term recurrence formulae of the ψ_j's

$$\beta_{n+2}(\psi'_n(\xi)\psi_{n+1}(t) - \psi'_{n+1}(\xi)\psi_n(t))$$
$$= \psi'_n(\xi)\left[(t - \alpha_{n+1})\psi_n(t) - \beta_{n+1}\psi_{n-1}(t)\right]$$
$$\quad - \psi_n(t)\left[(\xi - \alpha_{n+1})\psi'_n(\xi) + \psi_n(\xi) - \beta_{n+1}\psi'_{n-1}(\xi)\right]$$
$$= \beta_{n+1}(\psi'_{n-1}(\xi)\psi_n(t) - \psi'_n(\xi)\psi_{n-1}(t))$$
$$\quad + (t - \xi)\psi'_n(\xi)\psi_n(t) - \psi_n(t)\psi_n(\xi).$$

A repeated application of this step finally yields the wanted expression (2.6.5). □

Recurrence Relations for Hermite Kernel Polynomials

For practical purposes, the representation (2.6.10) of ψ_n^{ME} is not very useful, since in general all previous basis polynomials are required. In this section we work out short recursions for the scaled Hermite kernel polynomials.

We will start with bad news: The ψ_n^{ME}'s do **not** fulfill a three-term recurrence relation in the classical sense. Fortunately, however, it is possible to develop "mixed recurrence relations". In accordance with the kernel polynomial case (cf. (2.5.8) and (2.5.9)) we investigate two versions

(2.6.13) $$\psi_n^{\text{ME}}(t) = \psi_{n-1}^{\text{ME}}(t) - \rho_{n-1}t^2 p_{n-1}(t).$$

and

(2.6.14) $$\psi_n^{\text{ME}}(t) = \psi_{n-1}^{\text{ME}}(t) - \hat{\rho}_n t p_n(t).$$

Here, p_n and p_{n-1} are suitable chosen polynomials of exact degree n and $n-1$, respectively. Moreover, ρ_n and $\hat{\rho}_{n-1}$ are parameters to be determined.

It is worth noticing that each representation leads to a different, though mathematically equivalent, implementation of the ME scheme as described in Example 2.6.2. The first Ansatz leads to the OD (**O**rthogonal **D**irection) method of Fridman [53] and Fletcher [46]. In fact, the **ST**abilized version STOD, due Stoer and Freund [119], results. Finally, expression (2.6.14) (partly) generates the SYMMLQ implementation of Paige and Saunders [95].

To see that the ψ_n^{ME}'s do not fulfill a three-term recurrence relation, assume, to the contrary, that there exist recurrence coefficients with

$$\gamma_n \psi_n^{\mathrm{ME}}(t) = (t - \alpha_n)\psi_{n-1}^{\mathrm{ME}}(t) - \beta_n \psi_{n-2}^{\mathrm{ME}}(t).$$

Now the interpolatory constraints for the ψ_j^{ME}'s imply

$$0 = \gamma_n \frac{d\psi_n^{\mathrm{ME}}}{dt}(0) = \psi_{n-1}^{\mathrm{ME}}(0) = 1,$$

a contradiction.

Let us now turn to the representation (2.6.13), i.e., let us determine a polynomial p_{n-1} and a number ρ_{n-1} with

$$\psi_n^{\mathrm{ME}}(t) = \psi_{n-1}^{\mathrm{ME}}(t) - \rho_{n-1} t^2 p_{n-1}(t).$$

Obviously, by construction, both interpolatory constraints on ψ_n^{ME} are fulfilled for any choice of p_{n-1} and ρ_{n-1}, respectively. In view of Theorem 2.6.1, we have to adjust these degrees of freedom such that

(2.6.15) $\qquad\qquad \langle \psi_n^{\mathrm{ME}}, q \rangle = 0, \quad \text{for all} \quad q \in \Pi_{n-1}.$

Just as in the case of ordinary kernel polynomials (compare the derivation of Algorithm 2.5.6), we choose $p_{n-1}(t) = \psi_{n-1}^{\mathrm{MO}}(t)$ to be the monic orthogonal polynomial with respect to the inner product $\langle \cdot, \cdot t^2 \rangle$. This setting together with (2.6.15) uniquely defines the free parameter

(2.6.16) $\qquad\qquad \rho_{n-1} = \dfrac{\langle \psi_{n-1}^{\mathrm{ME}}, \psi_{n-1}^{\mathrm{MO}} \rangle}{\langle \psi_{n-1}^{\mathrm{MO}}, t^2 \psi_{n-1}^{\mathrm{MO}} \rangle}.$

Obviously, the denominator of (2.6.16) never vanishes. By applying Corollary 2.6.4, we see that the degenerate case $\psi_n'(0) = 0$ implies $\rho_{n-1} = 0$, an instance which does not affect the following algorithm. Notice, furthermore, that this algorithm and Algorithm 2.5.6, for the generation of ψ_n^{MR}, differ only in the computation of the "steplength" ρ_{n-1}.

2.6 Hermite Kernel Polynomials

Algorithm 2.6.6 *The residual polynomials $\{\psi_j^{\mathrm{ME}}\}_{j=0}^n$ and the monic polynomials $\{\psi_j^{\mathrm{MO}}\}_{j=0}^{n-1}$, orthogonal with respect to $\langle \cdot, \cdot t^2 \rangle$, may be computed by the following recursive procedure.*

Set
$$\psi_0^{\mathrm{ME}}(t) = 1, \quad \psi_{-1}^{\mathrm{MO}}(t) := 0, \quad \psi_0^{\mathrm{MO}}(t) = 1.$$

For $j = 1, 2, \ldots, n$ compute

$$\rho_{j-1} = \frac{\langle \psi_{j-1}^{\mathrm{ME}}, \psi_{j-1}^{\mathrm{MO}} \rangle}{\langle \psi_{j-1}^{\mathrm{MO}}, t^2 \psi_{j-1}^{\mathrm{MO}} \rangle}, \quad \psi_j^{\mathrm{ME}}(t) = \psi_{j-1}^{\mathrm{ME}}(t) - \rho_{j-1} t^2 \psi_{j-1}^{\mathrm{MO}}(t),$$

$$\alpha_j = \frac{\langle t\psi_{j-1}^{\mathrm{MO}}, t^2 \psi_{j-1}^{\mathrm{MO}} \rangle}{\langle \psi_{j-1}^{\mathrm{MO}}, t^2 \psi_{j-1}^{\mathrm{MO}} \rangle}, \quad \beta_j = \frac{\langle \psi_{j-1}^{\mathrm{MO}}, t^2 \psi_{j-1}^{\mathrm{MO}} \rangle}{\langle \psi_{j-2}^{\mathrm{MO}}, t^2 \psi_{j-2}^{\mathrm{MO}} \rangle}, \quad \beta_1 := 0,$$

$$\psi_j^{\mathrm{MO}}(t) = (t - \alpha_j)\psi_{j-1}^{\mathrm{MO}}(t) - \beta_j \psi_{j-2}^{\mathrm{MO}}(t).$$

Just as in the case of ordinary kernel polynomials (compare Theorem 2.5.7), the free parameter in the second mixed recurrence relation (2.6.14) is fixed using the QR factorization (2.3.2).

Theorem 2.6.7 *Let $\{\psi_n, \hat{J}_n; \langle \cdot, \cdot \rangle\}$ (cf. (2.1.30)) and the QR factorization $\{Q_n, R_n, G_n(c_n, s_n); \hat{J}_n\}$ (cf. (2.3.6)) be given. Furthermore, let ψ_n^{ME} denote the residual polynomial (2.6.11) and let $\hat{\eta}_n^{\mathrm{ME}}$ be as defined in Cor. 2.6.3. Then the following recurrence relations hold.*

$$\psi_n^{\mathrm{ME}}(t) = \psi_{n-1}^{\mathrm{ME}}(t) - \hat{\eta}_n^{\mathrm{ME}} t w_n^{\mathrm{ME}}(t)$$

where
$$w_n^{\mathrm{ME}}(t) = c_n w_n(t) + s_n \psi_n(t),$$

$$w_n(t) = -s_{n-1} w_{n-1}(t) + c_{n-1} \psi_{n-1}(t), \quad w_1(t) = \psi_0(t).$$

Proof. We first apply (2.3.2) and Cor. 2.6.3(a) to deduce an alternative representation for ψ_n^{ME}

$$\psi_n^{\text{ME}}(t) = 1 - t^2 \Psi_{n-1}^T(t) y_n^{\text{ME}}$$

(2.6.17)
$$= 1 - t\Psi_n^T(t) Q_n^T \begin{bmatrix} R_n \\ 0 \end{bmatrix} y_n^{\text{ME}}$$

$$= 1 - t\Psi_n^T(t) Q_n^T \begin{bmatrix} \hat{y}_n^{\text{ME}} \\ 0 \end{bmatrix}.$$

From this and Lemma 2.3.2(b) we have

$$\psi_n^{\text{ME}}(t) = 1 - t\Psi_n^T(t) Q_n^T \begin{bmatrix} \hat{y}_{n-1}^{\text{ME}} \\ \hat{\eta}_n^{\text{ME}} \\ 0 \end{bmatrix}$$

$$= 1 - t\Psi_{n-1}^T(t) Q_{n-1}^T \begin{bmatrix} \hat{y}_{n-1}^{\text{ME}} \\ 0 \end{bmatrix} - \hat{\eta}_n^{\text{ME}} t\Psi_n^T(t) Q_n^T \begin{bmatrix} e_n \\ 0 \end{bmatrix}$$

$$= \psi_{n-1}^{\text{ME}}(t) - \hat{\eta}_n^{\text{ME}} t\Psi_n^T(t) Q_n^T \begin{bmatrix} e_n \\ 0 \end{bmatrix}$$

$$= \psi_{n-1}^{\text{ME}}(t) - \hat{\eta}_n^{\text{ME}} t(c_n \Psi_{n-1}^T(t) Q_{n-1}^T e_n + s_n \psi_n(t)).$$

Finally, Lemma 2.3.2(c) (considered for $n-1$) yields the remaining recursion

(2.6.18)
$$w_n(t) := \Psi_{n-1}^T(t) Q_{n-1}^T e_n$$
$$= -s_{n-1} \Psi_{n-2}^T(t) Q_{n-2}^T e_{n-1} + c_{n-1} \psi_{n-1}(t)$$
$$= -s_{n-1} w_{n-1}(t) + c_{n-1} \psi_{n-1}(t).$$

□

We finish this section by once more discussing the degenerate case $\psi_n'(0) = 0$. Since c_n and s_n do not vanish together, the polynomial w_n^{ME} can not vanish identically $w_n^{\text{ME}} \not\equiv 0$. Hence, we learn from the previous theorem that $\psi_n^{\text{ME}} \equiv \psi_{n-1}^{\text{ME}}$ is only possible if $\hat{\eta}_n^{\text{ME}} = 0$. This, together with Cor. 2.6.4, proves the next result.

Corollary 2.6.8 *Let $\{\psi_n; \langle \cdot, \cdot \rangle\}$ (cf. (2.1.13)) be given. Furthermore, let $\hat{\eta}_n^{\text{ME}}$ be as defined in Cor. 2.6.3. Then*

$$\psi_n'(0) = 0 \quad \text{if, and only if} \quad \hat{\eta}_n^{\text{ME}} = 0.$$

2.6 Hermite Kernel Polynomials

Zeros of Hermite Kernel Polynomials

To discuss the zeros of $K_n^H(t;\xi)$ we make use of the Christoffel-Darboux formula as stated in Theorem 2.6.5

$$K_n^H(t;\xi) = \beta_{n+2}(\psi_n'(\xi)\psi_{n+1}(t) - \psi_{n+1}'(\xi)\psi_n(t)).$$

This relation ensures that $K_n^H(t;\xi)$ has simple zeros which interlace with the one of ψ_n and ψ_{n+1}, respectively.

Theorem 2.6.9 *Let $\{\psi_n;\langle\cdot,\cdot\rangle\}$ (cf. (2.1.13)) be given. Moreover, let $\{\theta_j^{(n)}\}_{j=1}^n$ and $\{\theta_j^{(n+1)}\}_{j=1}^{n+1}$ denote the zeros of ψ_n and ψ_{n+1}, respectively. Finally, define $\theta_0^{(n)} := -\infty$ and $\theta_{n+1}^{(n)} := \infty$.*

(a) *If $\psi_n'(\xi) = 0$, then $K_n^H(t;\xi)$ has the same zeros as ψ_n.*

(b) *If $\psi_{n+1}'(\xi) = 0$, then $K_n^H(t;\xi)$ has the same zeros as ψ_{n+1}.*

(c) *If $\psi_{n+1}'(\xi) \cdot \psi_n'(\xi) > 0$, then $K_n^H(t;\xi)$ has precisely one zero in each interval*

$$(\theta_j^{(n+1)}, \theta_j^{(n)}), \quad j = 1, 2, \ldots, n+1.$$

(d) *If $\psi_{n+1}'(\xi) \cdot \psi_n'(\xi) < 0$, then $K_n^H(t;\xi)$ has precisely one zero in each interval*

$$(\theta_{j-1}^{(n)}, \theta_j^{(n+1)}), \quad j = 1, 2, \ldots, n+1.$$

Proof. Part (a) and (b) are an immediate consequence of the Christoffel-Darboux formula (cf. Theorem 2.6.5). The proof of part (c) and (d), respectively, follows the lines of the proof of the corresponding parts of Theorem 2.5.8. Instead of discussing the sign distribution of $P_n(t;\xi)$, we investigate the polynomial

$$\hat{P}_n(t;\xi) := \psi_{n+1}(t) - \frac{\psi_{n+1}'(\xi)}{\psi_n'(\xi)}\psi_n(t)$$

which clearly has the same zeros as $\mathcal{K}_n(t;\xi)$. We have

$$\operatorname{sgn}\left(\hat{P}_n(\theta_j^{(n)};\xi)\right) = (-1)^{j+n+1},$$

$$\operatorname{sgn}\left(\hat{P}_n(\theta_j^{(n+1)};\xi)\right) = \begin{cases} (-1)^{j+n} & \text{if } \psi'_{n+1}(\xi)/\psi'_n(\xi) > 0, \\ (-1)^{j+n+1} & \text{if } \psi'_{n+1}(\xi)/\psi'_n(\xi) < 0, \end{cases}$$

and the wanted results follow. □

As shown in the preceding sections, the zeros of orthogonal polynomials and the zeros of kernel polynomials can be computed (in a stable fashion) by solving some appropriate eigenvalue problem. Unfortunately, such an algorithm for the computation of the zeros of Hermite kernel polynomials does not seem to exist.

2.7 Orthogonal and (Hermite) Kernel Polynomials

In the previous sections we exploited the QR factorization of $\hat{J}_n$ (see Theorem 2.5.7 and Theorem 2.6.7, respectively) to devise stable recurrences for the residual polynomials ψ_n^{MR} and ψ_n^{ME}, respectively. It is the goal of this section to show that, based on these recurrences, it is also possible to come up with stable schemes for the generation of the orthogonal residual polynomial characterized in Theorem 2.4.5

$$(2.7.1) \qquad \psi_n^{OR}(t) = 1 - t\Psi_{n-1}^T(t)y_n^{OR}$$

where y_n^{OR} solves

$$(2.7.2) \qquad J_n y = \sqrt{\nu_0}e_1.$$

The trick is to rewrite J_n in terms of $\hat{J}_n$ and then to apply the QR factorization to the latter matrix. To do so, we have two options (cf. (2.1.26))

$$(2.7.3) \qquad J_n = \hat{J}_n^T \begin{bmatrix} I_n \\ 0 \end{bmatrix} = [R_n^T \ 0]Q_n \begin{bmatrix} I_n \\ 0 \end{bmatrix} = R_n^T \begin{bmatrix} I_{n-1} & 0 \\ 0 & c_n \end{bmatrix} Q_{n-1},$$

and

$$(2.7.4) \qquad J_n = J_n^T = Q_{n-1}^T \begin{bmatrix} I_{n-1} & 0 \\ 0 & c_n \end{bmatrix} R_n.$$

Recall that by Lemma 2.4.4 the linear system (2.7.2) has a solution, if, and only if $c_n > 0$.

2.7 Orthogonal and (Hermite) Kernel Polynomials

The ME Connection

We will show that it is indeed possible to generate the residual polynomial ψ_n^{OR} -if existing- while computing the minimal error polynomial ψ_n^{ME}.

Some historical notes are appropriate here. They concern stable implementations of polynomial acceleration methods defined by ψ_n^{OR}. Clearly, the breakthrough came with the landmark paper by Paige and Saunders [95]. They, however, were only interested in the iterates produced by ψ_n^{OR}. They considered the iterates generated by ψ_n^{ME} only as auxiliary vectors. Freund [47] was the first who showed that the ME iterates are characterized by a variational property (cf. Example 2.6.2). This fact was rediscovered by Szyld and Widlund [123].

We start by substituting J_n in (2.7.2) by the expression obtained in (2.7.3)

$$R_n^T \begin{bmatrix} I_{n-1} & 0 \\ 0 & c_n \end{bmatrix} \hat{y}_n^{\text{OR}} = \sqrt{\nu_0} e_1, \quad y_n^{\text{OR}} = Q_{n-1}^T \hat{y}_n^{\text{OR}}.$$

Note that $\hat{y}_n^{\text{OR}}$ is closely connected to $\hat{y}_n^{\text{ME}}$, defined in Cor. 2.6.3(a),

$$\hat{y}_n^{\text{OR}} = \begin{bmatrix} I_{n-1} & 0 \\ 0 & 1/c_n \end{bmatrix} \hat{y}_n^{\text{ME}}.$$

This allows us to write

$$(2.7.5) \quad y_n^{\text{OR}} = Q_{n-1}^T \begin{bmatrix} I_{n-1} & 0 \\ 0 & 1/c_n \end{bmatrix} \begin{bmatrix} \hat{y}_{n-1}^{\text{ME}} \\ \hat{\eta}_n^{\text{ME}} \end{bmatrix} = Q_{n-1}^T \begin{bmatrix} \hat{y}_{n-1}^{\text{ME}} \\ \hat{\eta}_n^{\text{ME}}/c_n \end{bmatrix}.$$

Now we are in a position to derive the wanted relation. Using the representations (2.7.1),(2.6.17) and (2.6.18) we find

$$\begin{aligned}
\psi_n^{\text{OR}}(t) &= 1 - t\Psi_{n-1}^T(t) y_n^{\text{OR}} \\
&= 1 - t\Psi_{n-1}^T Q_{n-1}^T \begin{bmatrix} \hat{y}_{n-1}^{\text{ME}} \\ \hat{\eta}_n^{\text{ME}}/c_n \end{bmatrix} \\
&= 1 - t\Psi_{n-1}^T Q_{n-1}^T \begin{bmatrix} \hat{y}_{n-1}^{\text{ME}} \\ 0 \end{bmatrix} - \frac{\hat{\eta}_n^{\text{ME}}}{c_n} t\Psi_{n-1}^T Q_{n-1}^T e_n \\
&= \psi_{n-1}^{\text{ME}}(t) - \frac{\hat{\eta}_n^{\text{ME}}}{c_n} t w_n(t).
\end{aligned}$$

Also, we can deduce from (2.7.5) an expression for the norm of ψ_n^{OR}. This is the second part of

Theorem 2.7.1 Let $\{\psi_n, (\alpha_n, \beta_n), \hat{J}_n, ; \langle \cdot, \cdot \rangle\}$ (cf. (2.1.30)) and the QR factorization $\{Q_n, R_n, G_n(c_n, s_n); \hat{J}_n\}$ (cf. (2.3.6)) be given. Furthermore, let ψ_n^{OR} and ψ_{n-1}^{ME} denote the residual polynomials (2.7.1) and (2.6.11), respectively. Finally, let $\hat{\eta}_n^{\mathrm{ME}}$ and w_n be as defined in Cor. 2.6.3 and Theorem 2.6.7, respectively. If $c_n \neq 0$, we have

(a) $$\psi_n^{\mathrm{OR}}(t) = \psi_{n-1}^{\mathrm{ME}}(t) - \frac{\hat{\eta}_n^{\mathrm{ME}}}{c_n} t w_n(t).$$

(b) $$\|\psi_n^{\mathrm{OR}}\| = \beta_{n+1} \left| s_{n-1} \hat{\eta}_{n-1}^{\mathrm{ME}} + \frac{c_{n-1}}{c_n} \hat{\eta}_n^{\mathrm{ME}} \right|.$$

Proof. (b) We apply (2.1.27) and Theorem 2.4.5 to obtain

(2.7.6)
$$\begin{aligned}\psi_n^{\mathrm{OR}}(t) &= 1 - t\Psi_{n-1}^T(t) y_n^{\mathrm{OR}} \\ &= 1 - \Psi_{n-1}^T(t) J_n y_n^{\mathrm{OR}} - \beta_{n+1} \psi_n(t) e_n^T y_n^{\mathrm{OR}} \\ &= -\beta_{n+1} \psi_n(t) e_n^T y_n^{\mathrm{OR}}.\end{aligned}$$

Hence, we need to compute the last component of y_n^{OR}. With (2.7.5), (2.3.4), and Cor. 2.6.3(a) it follows

$$y_n^{\mathrm{OR}} = Q_{n-1}^T \begin{bmatrix} \hat{y}_{n-1}^{\mathrm{ME}} \\ \hat{\eta}_n^{\mathrm{ME}}/c_n \end{bmatrix} = \begin{bmatrix} Q_{n-2}^T & 0 \\ 0 & 1 \end{bmatrix} G_{n-1}^T \begin{bmatrix} \hat{y}_{n-2}^{\mathrm{ME}} \\ \hat{\eta}_{n-1}^{\mathrm{ME}} \\ \hat{\eta}_n^{\mathrm{ME}}/c_n \end{bmatrix},$$

which shows that

$$e_n^T y_n^{\mathrm{OR}} = s_{n-1} \hat{\eta}_{n-1}^{\mathrm{ME}} + \frac{c_{n-1}}{c_n} \hat{\eta}_n^{\mathrm{ME}}$$

and concludes the proof. □

We mention that, in contrast to the recursion based on Algorithm 2.4.2, the above recursion formula for the ψ_n^{OR} does not rely on previous ψ_j^{OR}, $j < n$, which may not exist.

2.7 Orthogonal and (Hermite) Kernel Polynomials

The MR Connection

Now we will derive the relationship between ψ_n^{OR} and ψ_n^{MR}. This time we rewrite J_n by means of the expression (2.7.4)

$$Q_{n-1}^T \begin{bmatrix} I_{n-1} & 0 \\ 0 & c_n \end{bmatrix} \hat{y}_n^{\text{OR}} = \sqrt{\nu_0} e_1, \quad y_n^{\text{OR}} = R_n^{-1} \hat{y}_n^{\text{OR}}.$$

By construction, the quantity $\hat{y}_n^{\text{OR}}$ looks almost like the corresponding one in the MR scheme (cf. Cor. 2.5.3)

$$\hat{y}_n^{\text{OR}} = \begin{bmatrix} I_{n-1} & 0 \\ 0 & 1/c_n \end{bmatrix} \begin{bmatrix} \hat{y}_{n-1}^{\text{MR}} \\ \hat{\eta}_{n-1}^{\text{MR}} \end{bmatrix}.$$

Consequently, we have

(2.7.7) $$y_n^{\text{OR}} = R_n^{-1} \begin{bmatrix} \hat{y}_{n-1}^{\text{MR}} \\ \hat{\eta}_{n-1}^{\text{MR}}/c_n \end{bmatrix}.$$

Using, in addition, (2.7.1), (2.1.28), (2.5.12), Theorem 2.5.7(b), and Corollary 2.5.3(a) it follows

(2.7.8)
$$\begin{aligned}
\psi_n^{\text{OR}}(t) &= 1 - t\Psi_{n-1}^T(t) y_n^{\text{OR}} \\
&= 1 - t\Psi_{n-1}^T(t) R_n^{-1} \begin{bmatrix} \hat{y}_{n-1}^{\text{MR}} \\ \hat{\eta}_{n-1}^{\text{MR}}/c_n \end{bmatrix} \\
&= 1 - \Psi_n^T(t) Q_n^T \begin{bmatrix} \hat{y}_{n-1}^{\text{MR}} \\ \hat{\eta}_{n-1}^{\text{MR}}/c_n \\ 0 \end{bmatrix} \\
&= 1 - \Psi_n^T(t) Q_n^T \begin{bmatrix} \hat{y}_n^{\text{MR}} \\ 0 \end{bmatrix} - \frac{s_n^2}{c_n} \hat{\eta}_{n-1}^{\text{MR}} \Psi_n^T(t) Q_n^T \begin{bmatrix} e_n \\ 0 \end{bmatrix} \\
&= \psi_n^{\text{MR}}(t) - \frac{s_n^2}{c_n} \hat{\eta}_{n-1}^{\text{MR}} t\Psi_{n-1}^T(t) R_n^{-1} e_n \\
&= \psi_n^{\text{MR}}(t) - \frac{s_n^2}{c_n} \hat{\eta}_{n-1}^{\text{MR}} t w_n^{\text{MR}}(t).
\end{aligned}$$

The obtained expression for ψ_n^{OR} is surprisingly close to the one for ψ_n^{MR} (cf. Theorem 2.5.7(a))

(2.7.9) $$\psi_n^{\text{MR}}(t) = \psi_{n-1}^{\text{MR}}(t) - c_n \eta_{n-1}^{\text{MR}} t w_n^{\text{MR}}(t).$$

Combining (2.7.8) and (2.7.9) leads to the remarkable fact, that ψ_n^{MR} is a convex combination of ψ_n^{OR} and ψ_{n-1}^{MR}. Once again it is possible to come up with a computable expression for the norm of ψ_n^{OR}. Moreover, we offer an alternative expression for the Givens parameter c_n (compare Cullum and Greenbaum [25, Theorem 6]). Altogether, we have

---

Theorem 2.7.2 *Let $\{\psi_n, (\alpha_n, \beta_n), \hat{J}_n, ; \langle \cdot, \cdot \rangle\}$ (cf. (2.1.30)) and the QR factorization $\{Q_n, R_n, G_n(c_n, s_n); \tilde{J}_n\}$ (cf. (2.3.6)) be given. Furthermore, let ψ_n^{OR} and ψ_{n-1}^{MR} denote the residual polynomials (2.7.1) and (2.5.5), respectively. Finally, let $\hat{\eta}_n^{\mathrm{MR}}$ and w_n^{MR} be as defined in Cor. 2.5.3 and Theorem 2.5.7, respectively. If $c_n \neq 0$, we have*

(a) $$\psi_n^{\mathrm{OR}}(t) = \psi_n^{\mathrm{MR}}(t) - \frac{s_n^2}{c_n}\hat{\eta}_{n-1}^{\mathrm{MR}} t w_n^{\mathrm{MR}}(t).$$

(b) $$\psi_n^{\mathrm{MR}}(t) = c_n^2 \psi_n^{\mathrm{OR}}(t) + s_n^2 \psi_{n-1}^{\mathrm{MR}}(t).$$

(c) $$\|\psi_n^{\mathrm{OR}}\| = \frac{\sqrt{\nu_0}}{c_n}\prod_{j=1}^n |s_j| = \frac{1}{c_n}\|\psi_n^{\mathrm{MR}}\|.$$

(d) $$c_n^2 = 1 - \frac{\|\psi_n^{\mathrm{MR}}\|^2}{\|\psi_{n-1}^{\mathrm{MR}}\|^2}.$$

Proof. To compute the norm of ψ_n^{OR} we start with (2.7.6) and we subsequently apply (2.7.7) and Lemma 2.3.1 to obtain

$$\psi_n^{\mathrm{OR}}(t) = -\beta_{n+1}\psi_n(t)e_n^T y_n^{\mathrm{OR}} = -\frac{\beta_{n+1}\hat{\eta}_{n-1}^{\mathrm{MR}}}{c_n r_{1,n}}\psi_n(t) = -\frac{s_n \hat{\eta}_{n-1}^{\mathrm{MR}}}{c_n}\psi_n(t).$$

Hence $\|\psi_n^{\mathrm{OR}}\|$ coincides with $\|\psi_n^{\mathrm{MR}}\|$ up to the quotient c_n (cf. Cor. 2.5.3). Finally, part (d)

$$\|\psi_n^{\mathrm{MR}}\| = s_n\|\psi_{n-1}^{\mathrm{MR}}\| = \sqrt{1-c_n^2}\|\psi_{n-1}^{\mathrm{MR}}\|$$

follows directly from Cor. 2.5.3(b). $\square$

2.7 Orthogonal and (Hermite) Kernel Polynomials

The ME - MR Connection

We finish this section by noting that it is possible to compute minimal residual polynomials out of the minimal error polynomials. To show this, we only have to combine the two previous theorems.

Theorem 2.7.3 Let $\{\psi_n, \hat{J}_n, ; \langle \cdot, \cdot \rangle\}$ (cf. (2.1.30)) and the QR factorization $\{Q_n, R_n, G_n(c_n, s_n); \hat{J}_n\}$ (cf. (2.3.6)) be given. Furthermore, let ψ_n^{MR} and ψ_n^{ME} denote the residual polynomials (2.7.1) and (2.5.5), respectively. Finally, let $\hat{\eta}_n^{\mathrm{ME}}$ and w_n be as defined in Cor. 2.6.3 and Theorem 2.6.7, respectively. Then the following recursion holds

$$\psi_n^{\mathrm{MR}}(t) = s_n^2 \psi_{n-1}^{\mathrm{MR}}(t) + c_n^2 \psi_{n-1}^{\mathrm{ME}}(t) - c_n \hat{\eta}_n^{\mathrm{ME}} t w_n(t).$$

---

3 Chebyshev and Optimal Polynomials

In this chapter we outline how to compute Chebyshev polynomials and certain closely related optimal polynomials for one interval and for the union of two disjoint intervals, respectively.

3.1 Basic Definitions

Let A be a real symmetric matrix and let $\Omega \supset \sigma(A)$ denote a set that contains all eigenvalues of A. More precisely, we assume that

$$\sigma(A) \subset \Omega = [a,d], \quad 0 < a < d,$$

for positive definite A and that

(3.1.1) $\qquad \sigma(A) \subset \Omega = E := [a,b] \cup [c,d] \quad a < b < 0 < c < d,$

for indefinite A. Now the question arises: what use can be made out of the "information Ω" ?

To start with, consider the polynomial based method

$$r_n = p_n(A) r_0, \quad p_n(0) = 1.$$

We may estimate the Euclidian norm of the residual

$$\|r_n\|_2 = \|p_n(A) r_0\|_2 \leq \max_{\lambda \in \sigma(A)} |p_n(\lambda)| \, \|r_0\|_2 \leq \max_{t \in \Omega} |p_n(t)| \, \|r_0\|_2$$

in terms of the *Chebyshev norm* over Ω

(3.1.2) $\qquad \dfrac{\|r_n\|_2}{\|r_0\|_2} \leq \|p_n\|_\Omega, \quad \text{where} \quad \|p\|_\Omega := \max_{t \in \Omega} |p(t)|.$

Clearly, an obvious choice for the residual polynomial p_n is the one which minimizes the upper bound in (3.1.2). This *optimal polynomial* $\mathcal{P}_n(t; \Omega, 0)$ is the solution of the following Chebyshev approximation problem

(3.1.3) $\qquad \|\mathcal{P}_n(t; \Omega, 0)\|_\Omega = \min \{ \|p\|_\Omega : p \in \Pi_n, \, p(0) = 1 \}, \quad 0 \notin \Omega.$

3.1 Basic Definitions

The point 0, arising in the interpolatory constraint $p(0) = 1$, will be called *constraint point*. As it is not surprising, optimal polynomials are closely connected to the solution $\mathcal{T}_n(t;\Omega)$ of the classical Chebyshev approximation problem

$$(3.1.4) \qquad \|\mathcal{T}_n(t;\Omega)\|_\Omega = \min\left\{\|t^n - p(t)\|_\Omega : p \in \Pi_{n-1}\right\}.$$

$\mathcal{T}_n(t;\Omega)$ is called *Chebyshev polynomials with respect to* Ω. Note that the scaled Chebyshev polynomial $\mathcal{T}_n(t;\Omega)/\mathcal{T}_n(0;\Omega)$ should be a "good candidate" for the optimal polynomial with respect to Ω. We remark that standard results from approximation theory (cf. Meinardus[91]) guarantee that there always exists a unique solution to each of the above Chebyshev approximation problems. The characterization and the actual computation of the best approximations will be postponed to a later section.

There is yet another fact which motivates us to have a closer look at the optimal polynomials. To measure the performance of a particular scheme $\varepsilon_n = p_n(A)\varepsilon_0$ we define the *root-convergence factor* $\kappa(A;\{p_n\})$ (cf. Eiermann and Niethammer [34]) by

$$\kappa(A;\{p_n\}) := \limsup_{n\to\infty} \left(\sup_{\varepsilon_0 \neq 0} \frac{\|\varepsilon_n\|_2}{\|\varepsilon_0\|_2}\right)^{1/n}.$$

It is also called the *asymptotic convergence factor* for the polynomial iteration method induced by $\{p_n\}$. The *asymptotic convergence factor for* Ω is then defined by

$$\kappa(\Omega) := \inf_{\{p_n\}} \left\{\sup\left(\kappa(A;\{p_n\}): A \in \mathbb{R}^{N\times N},\ N \geq 1,\ \text{and } \sigma(A) \subset \Omega\right)\right\}.$$

A scheme, induced by $\{p_n\}$, with the property

$$\kappa(A;\{p_n\}) \leq \kappa(\Omega),$$

for any A with $\sigma(A) \subset \Omega$, will be called *asymptotically optimal with respect to* Ω.

It is intuitively clear that the scheme induced by the optimal polynomial (3.1.3) is asymptotically optimal. Moreover, Eiermann, Niethammer and Varga [35] devised the following alternate characterization

$$(3.1.5) \qquad \kappa(\Omega) = \lim_{n\to\infty} \left(\|\mathcal{P}_n(t;\Omega,0)\|_\Omega\right)^{1/n}.$$

This expression, however, is still not well suited for numerical purposes. Here, Eiermann, Li and Varga [33] came up with a beautiful representation of $\kappa(\Omega)$ in terms of the Green's function $g(z; \Omega^c, \infty)$ for the complement Ω^c of Ω with pole at infinity

$$(3.1.6) \qquad \kappa(\Omega) = \exp(-g(0; \Omega^c, \infty)).$$

In the next sections we will explicitly compute the optimal polynomials for one interval and for the union of two disjoint intervals. Surprisingly enough, these polynomials enjoy a representation in terms of the associated Green's function. So, we obtain, as a by-product, a computable formula for the corresponding convergence factor $\kappa(\Omega)$. Moreover, it turns out that for the cases under consideration the optimal polynomials (3.1.3) are just scaled Chebyshev polynomials (3.1.4). Finally, and most important, we will show that the optimal polynomials fulfill a three-term recurrence relation and thus lead to very attractive iteration schemes..

Green's Function

In this section we introduce the Green's function and collect some basic properties (compare Widom [134]).

Let $\overline{\mathbb{C}} := \mathbb{C} \cup \infty$ denote the extended complex plane and let $\Omega^c := \overline{\mathbb{C}} \setminus \Omega$ denote the complement of Ω with respect to $\overline{\mathbb{C}}$.

The existence of a *Green's function* $g(z; \Omega^c, z_0)$ for Ω^c with pole at $z_0 \in \Omega^c$ is, for example, guaranteed if Ω^c is of finite connectivity and if it is the complement of a compact set which has no isolated points (cf. Walsh [127, pp. 65]). Moreover, the Green's function $g(z; \Omega^c, z_0)$ is uniquely characterized by the following three properties:

(a) $g(z; \Omega^c, z_0)$ is real harmonic in Ω^c except at z_0,

(b) the difference function

$$g(z; \Omega^c, z_0) - \begin{cases} \log \dfrac{1}{|z - z_0|} & \text{if } z_0 \neq \infty \\ \log |z| & \text{if } z_0 = \infty \end{cases}$$

is harmonic in a neighborhood of z_0,

3.1 Basic Definitions

(c) $g(z; \Omega^c, z_0) \to 0$ as $z \to \partial\Omega^c$, for $z \in \Omega^c$, where $\partial\Omega^c$ denotes the boundary of Ω^c.

We mention that the (positive) real valued Green's function $g(z; \Omega^c, z_0)$ may also be expressed in terms of a complex valued function $\mathcal{G}(z; \Omega^c, z_0)$. To see this, let $\mathcal{G}(z; \Omega^c, z_0)$ have the following properties

(a) $\mathcal{G}(z; \Omega^c, z_0)$ is an analytic function in Ω^c with a single valued modulus $|\mathcal{G}(z; \Omega^c, z_0)| < 1$ in Ω^c,

(b) $\mathcal{G}(z; \Omega^c, z_0)$ has precisely one simple zero $z_0 \in \Omega^c$,

(c) $|\mathcal{G}(z; \Omega^c, z_0)| = 1$, for $z \in \partial\Omega^c$.

We remark that $\mathcal{G}(z; \Omega^c, z_0)$ is not uniquely determined by these requirements (compare Example 3.1.1(a)). However, it is easy to see that

$$(3.1.7) \quad g(z; \Omega^c, z_0) = -\operatorname{Re}\bigl[\log(\mathcal{G}(z; \Omega^c, z_0))\bigr] = -\log(|\mathcal{G}(z; \Omega^c, z_0)|).$$

Therefore, the function $\mathcal{G}(z; \Omega^c, z_0)$ is known as *complex Green's function* (cf. Achieser [1, p. 1178]). Vice versa, if $h(z; \Omega^c, z_0)$ denotes a harmonic conjugate of $g(z; \Omega^c, z_0)$, then $\mathcal{G}(z; \Omega^c, z_0) := \exp(-g(z; \Omega^c, z_0) + ih(z; \Omega^c, z_0))$ is a complex Green's function.

For the construction of the Green's function for a particular set the method of conformal transplantation (compare Example 3.1.1) is often useful. If two sets are conformally equivalent, then the Green's function with corresponding poles are equal at points which correspond to each other (cf. Ahlfors [4, p. 249]).

In the following example we will compute the Green's function for the unit disc $D := \{v \in \mathbb{C} : |v| < 1\}$ and for the complement of the unit interval $I^c := [-1, 1]^c$. In particular, we will apply the method of conformal transplantation. Before we do so, let us mention that the so-called *Joukowsky map*

$$(3.1.8) \quad z = \phi_\mathrm{I}(v) := \frac{1}{2}\left(v + \frac{1}{v}\right)$$

provides the conformal mapping from the unit disc onto the complement of the unit interval $\phi_\mathrm{I}(D) = I^c$ (cf. Henrici [79, p. 297]). The inverse function is given by

$$v = \phi_\mathrm{I}^{-1}(z) = z - \sqrt{z^2 - 1},$$

where that value of the square root is to be chosen for which $|v| = |\phi_I^{-1}(z)| < 1$. We mention that, in view of the Riemann mapping theorem (cf. Ahlfors [4, p. 222]), the unit disc may be viewed as the canonical region for simply connected regions.

Example 3.1.1 (a) For any $\tau \in \mathbb{R}$ the function $\mathcal{G}(v; D, 0) := v \exp(i\tau)$ is a complex Green's function for the unit disc D. Consequently, we obtain

$$g(v; D, 0) = -\log |\mathcal{G}(v; D, 0)| = -\log |v|.$$

(b) To compute the Green's function with respect to I^c and pole at infinity, we apply (a) and the method of conformal transplantation

$$g(z; I^c, \infty) = g(\phi_I^{-1}(z); \phi_I^{-1}(I^c), \phi_I^{-1}(\infty)) = g(z - \sqrt{z^2 - 1}; D, 0)$$
$$= -\log |z - \sqrt{z^2 - 1}| = \log |z + \sqrt{z^2 - 1}|.$$

Notice further that $\mathcal{G}(z; I^c, \infty) := \phi_I^{-1}(z)$ is a complex Green's function for I^c. □

There is a close connection between the Green's function for Ω^c and the *equilibrium distribution for* Ω. Since this special distribution may be "visualized" as the solution of a physical problem, we will briefly outline this connection.

Equilibrium Distribution

The equilibrium distribution is one of the basic ingredients of *potential theory*. An introduction to potential theory can be found, for example, in Helms [78] and in Hille [82].

The terminology in potential theory arises from an electrostatics problem. To keep things easy, let $\Omega \subset \mathbb{C}$ be a compact set with no isolated points, such that the complement is of finite connectivity. Then, place a unit positive charge on Ω so that the equilibrium is reached in the sense that the energy with respect to the logarithmic potential is minimized.

To create a mathematical framework for this problem, consider the logarithmic potential with respect to the measure ν

$$U(z; \nu) := \int_\Omega \log \frac{1}{|z - t|} d\nu(t).$$

3.1 Basic Definitions

Then its energy is defined by

$$E(\nu) := \int_\Omega U(z;\nu) d\nu(z) = \iint_\Omega \log \frac{1}{|z-t|} d\nu(t) d\nu(z).$$

Thus, the electrostatics problem involves the determination of

$$V(\Omega) := \inf \{E(\nu) : \nu \text{ unit Borel measure supported on } \Omega\}.$$

A well-known result in potential theory (cf. Hille [82, Theorem 16.4.3]) asserts that this problem has a unique solution

$$E(\nu_\Omega) = V(\Omega).$$

The measure ν_Ω is the *equilibrium distribution for* Ω and furnishes the solution of the electrostatics problem. Moreover, the Green's function with respect to Ω^c admits the representation (cf. Hille [82, §16.5])

$$g(z; \Omega^c, \infty) = -U(z; \nu_\Omega) + E(\nu_\Omega).$$

This fact together with Example 3.1.1(b) gives the equilibrium distribution for the unit interval.

Example 3.1.2 Let $\Omega = [-1, 1]$. Then the equilibrium distribution

$$d\nu_\Omega(t) = \frac{1}{\pi} \frac{dt}{\sqrt{1-t^2}}, \quad t \in [-1, 1].$$

is just the arcsine measure (compare Example 2.1.2). Hence, the classical Chebyshev polynomials of the first kind are orthogonal with respect to the equilibrium distribution. □

To decide whether a particular polynomial is the optimal and/or the Chebyshev polynomial one needs a sufficient and necessary characterization for best approximations.

Characterization of the Best Approximation

Let Ω be a compact subset of the real line $\mathbb{R}$ and let g be a continuous function on Ω. In this section we characterize the best approximation of g with respect to the set of all polynomials that have degree not larger than $n-1$ and that satisfy some interpolatory constraints. More precisely, for a given number $0 \leq l < n-1$ and for given distinct knots $a_1 < a_2 < \cdots < a_l$ satisfying $a_j \notin \Omega$, and given numbers $A_j \in \mathbb{R}$, define

$$(3.1.9) \qquad \Pi_{n-1}^l(a_j, A_j) := \{p \in \Pi_{n-1} : p(a_j) = A_j, j = 1, 2, \ldots, l\}$$

Obviously, the dimension of this affine space is $\dim(\Pi_{n-1}^l) = n-l$. We are looking for the *best uniform approximation* p_{n-1} on Ω to g out of $\Pi_{n-1}^l(a_j, A_j)$, i.e.,

$$(3.1.10) \qquad \|g - p_{n-1}\|_\Omega = \min\left\{\|g - p\|_\Omega : p \in \Pi_{n-1}^l(a_j, A_j)\right\}.$$

It is straightforward to verify that (3.1.10) has a unique solution.

Let us now study the characteristic properties of the best approximation. The proof goes back to Achieser [1] for the special case $\Omega = E$ (cf. (3.1.1)). Almost 50 years later it was generalized by Grcar[72]. Nevertheless, the characterization seems to be not well known except to experts.

In almost any characterization theorem for Chebyshev approximation problems, the *extremal points* of the error function $g - p_{n-1}$

$$\mathrm{Ext}(g - p_{n-1}; \Omega) := \{t \in \Omega : |(g - p_{n-1})(t)| = \|g - p_{n-1}\|_\Omega\}$$

play a crucial role.

Theorem 3.1.3 *Let Ω be a given subset of the real line that contains at least $n - l + 1$, $0 \leq l < n - 1$, points. Furthermore, let g be a continuous function on Ω and let $\Pi_{n-1}^l(a_j, A_j)$ be defined as in (3.1.9).*

Then, p_{n-1} is the best approximation on Ω to g out of $\Pi_{n-1}^l(a_j, A_j)$ if, and only if the error function $g - p_{n-1}$ has $n - l + 1$ extremal points $t_1 < t_2 < \cdots < t_{n-l+1} \in \mathrm{Ext}(g - p_{n-1}; \Omega)$ with

$$(g - p_{n-1})(t_j) = (-1)^{\tau_j + 1}(g - p_{n-1})(t_{j+1}), \quad j = 1, 2, \ldots, n - l,$$

where τ_j denotes the number of knots $a_i \in (t_j, t_{j+1})$.

3.2 Chebyshev and Optimal Polynomials; One Interval

From the theorem above it is a small step to the characterization of optimal and Chebyshev polynomials, respectively.

Corollary 3.1.4 *Let Ω be a given subset of the real line that contains at least n points.*

(a) Let $\xi \notin \Omega$ be given. Then, $\mathcal{P}_n(t; \Omega, \xi)$ is the optimal polynomial for Ω with respect to ξ (cf. (3.1.3)) if, and only if there exist $n+1$ extremal points $t_1 < t_2 < \cdots < t_{n+1} \in Ext(\mathcal{P}_n; \Omega)$ with

$$\mathcal{P}_n(t_j; \Omega, \xi) = (-1)^{\tau+1} \mathcal{P}_n(t_{j+1}; \Omega, \xi),$$

where $\tau = 1$ if $\xi \in [t_j, t_{j+1}]$ and $\tau = 0$ otherwise.

(b) $\mathcal{T}_n(t; \Omega)$ is the Chebyshev polynomial for Ω (cf. (3.1.4)) if, and only if there exist $n+1$ extremal points $t_1 < t_2 < \cdots < t_{n+1} \in Ext(\mathcal{T}_n; \Omega)$ with

$$\mathcal{T}_n(t_j; \Omega) = -\mathcal{T}_n(t_{j+1}; \Omega).$$

That is, the t_j's form an alternating set for $\mathcal{T}_n(t; \Omega)$.

3.2 Chebyshev and Optimal Polynomials; One Interval

In this section we will compute the optimal and the Chebyshev polynomials (cf. (3.1.4) and (3.1.3)) for the "classical case" $\Omega = [a, d]$, $0 < a < d$. First, we will show that the optimal polynomials are just scaled Chebyshev polynomials. Then we will derive an explicit expression for the Chebyshev polynomial and work out its relation to the associated Green's function, Finally, we will investigate the orthogonality properties of the Chebyshev polynomials. These are all well known facts. However, their derivation will guide us in the far more "delicate case" of Ω being the union of two disjoint intervals.

It is convenient to first "normalize" the set Ω. Consider the linear transformation

$$(3.2.1) \qquad l(t) := \frac{a+d-2t}{a-d},$$

which maps $[a,d]$ onto the unit interval $I := [-1,1]$. In particular, the origin is mapped onto

$$(3.2.2) \qquad \xi := l(0) = \frac{a+d}{a-d}.$$

The Chebyshev approximation problem (3.1.4) is not affected by this linear transformation, that is

$$\mathcal{T}_n(t;[a,d]) = \left(\frac{d-a}{2}\right)^n \mathcal{T}_n(l(t);I).$$

Optimal polynomials are not invariant under linear transformations. Here, the constraint point makes the difference

$$\mathcal{P}_n(t;[a,d],0) = \mathcal{P}_n(l(t);I,\xi).$$

So, it suffices to study both approximation problems on the unit interval I, but (3.1.3) has to be investigated for arbitrary constraint points $\xi \notin I$.

Since the constraint point ξ is not located in the unit interval, we learn from Cor. 3.1.4, that both $\mathcal{T}_n(l(t);I)$ and $\mathcal{P}_n(l(t);I,\xi)$ are characterized by an alternating set consisting out of $n+1$ extremal points. Consequently, the optimal polynomial has to be the suitable scaled Chebyshev polynomial (cf. Markoff[90])

$$(3.2.3) \qquad \mathcal{P}_n(t;I,\xi) = \frac{\mathcal{T}_n(t;I)}{\mathcal{T}_n(\xi;I)}, \quad \xi \notin I.$$

Due to this identity we will in the remaining part of this section only consider Chebyshev polynomials. Notice that all zeros of $\mathcal{T}_n(t;I)$ are located in I and hence $\mathcal{T}_n(\xi;I) \neq 0$.

For convenience let us drop the dependence on the unit interval, i.e., we set

$$\mathcal{T}_n(t) = \mathcal{T}_n(t;I) \quad \text{and} \quad \mathcal{P}_n(t;\xi) = \mathcal{P}_n(t;I,\xi)$$

3.2 Chebyshev and Optimal Polynomials; One Interval

for the remaining part of this section.

In his original proof, back in 1859, Chebyshev[20] computed $\mathcal{T}_n$ as the solution of a certain differential equation. Let $L_n := \|\mathcal{T}_n\|_1$ denote the uniform norm of $\mathcal{T}_n$ on I. Since $\mathcal{T}_n$ has $n+1$ extremal points, it is easy to see that the two boundary points ± 1 have to be extremal points, whereas the remaining $n-1$ extremal points are relative maxima of $\mathcal{T}_n$, i.e., they are the zeros of $\mathcal{T}_n'$. Thus, the following equation holds

$$(3.2.4) \qquad (1-t^2)(\mathcal{T}_n'(t))^2 = n^2(L_n^2 - \mathcal{T}_n^2(t)),$$

or, equivalently, with $\mathcal{U}_{n-1}(t) := \mathcal{T}_n'(t)/n$,

$$(3.2.5) \qquad \mathcal{T}_n^2(t) + (1-t^2)\mathcal{U}_{n-1}^2(t) = L_n^2.$$

It turns out that (3.2.4) and (3.2.5) constitute key equations for the "unit interval case". To start with, we rewrite (3.2.4) as

$$(3.2.6) \qquad \int \frac{\mathcal{T}_n'(t)}{\sqrt{L_n^2 - \mathcal{T}_n^2(t)}} dt = n \int \frac{dt}{\sqrt{1-t^2}}.$$

To solve this equation substitute $t = \cos(t)$ and make use of the condition $\mathcal{T}_n(1) = L_n$ which then implies that $\mathcal{T}_n(t) = L_n \cos(n \arccos(t))$. Using trigonometric identities it is straightforward to verify, that $\cos(nx)$ is a polynomial of degree n in $t = \cos(x)$ and that its leading coefficient is 2^{n-1}. This leads to the familiar expression

$$(3.2.7) \qquad \mathcal{T}_n(t) = \frac{1}{2^{n-1}} \cos(n \arccos(t)).$$

On the other hand, one may as well perform a complex substitution in (3.2.6)

$$t = \cos(t) = \frac{1}{2}(e^{it} + e^{-it}) = \frac{1}{2}\left(v + \frac{1}{v}\right)$$

to eventually obtain a representation in terms of the Joukowsky map (3.1.8)

$$(3.2.8) \qquad \mathcal{T}_n(t) = \frac{1}{2^n}\left(v^n + \frac{1}{v^n}\right), \quad t = \frac{1}{2}\left(v + \frac{1}{v}\right).$$

This is quite an interesting representation. The next corollary provides a rather academic version of the expression (3.2.8).

> **Corollary 3.2.1** Let ϕ_I denote the conformal mapping from the unit disc D onto the complement I^c of the unit interval (cf. 3.1.8). Furthermore, let $\mathcal{G}(v; D, 0) := v$ denote a complex Green's function for the unit disc (cf. Example 3.1.1). Then, the Chebyshev polynomial for the unit interval may be represented as
>
> $$\mathcal{T}_n(t; I) = \frac{1}{2^n}\left(\mathcal{G}(v; D, 0)^n + \frac{1}{\mathcal{G}(v; D, 0)^n}\right), \quad t = \phi_I(v).$$

Based on the representation (3.2.7) one could now go ahead and work out orthogonality relations for $\mathcal{T}_n$. We will not follow these lines. Instead we will apply Theorem 2.2.8. This approach has the advantage in that it also works in more general situations, in particular in the case of two intervals.

After dividing both sides of (3.2.5) by $\mathcal{T}_n(t)^2(1-t^2)$ and after taking the square root, we obtain

$$\frac{\mathcal{U}_{n-1}(t)}{\mathcal{T}_n(t)} = \frac{1}{\sqrt{t^2-1}}\sqrt{1-\frac{L_n^2}{\mathcal{T}_n^2(t)}} = \frac{1}{\sqrt{t^2-1}}\left(1 - \frac{1}{2}\frac{L_n^2}{\mathcal{T}_n^2(t)} - \frac{1}{8}\frac{L_n^4}{\mathcal{T}_n^4(t)} - \cdots\right)$$

and consequently

$$(3.2.9) \qquad \frac{\mathcal{U}_{n-1}(t)}{\mathcal{T}_n(t)} = \frac{1}{\sqrt{t^2-1}} + O\left(\frac{1}{t^{2n+1}}\right), \quad \text{for } |t| \to \infty.$$

With the help of Cauchy's integral formula (cf. Ahlfors [4, p. 118]) we obtain, for $t \notin I$,

$$(3.2.10) \qquad \int_{-1}^{1} \frac{w(x)}{t-x}dx = \frac{1}{\sqrt{t^2-1}}, \quad \text{where } w(x) = \frac{1}{\sqrt{1-x^2}}.$$

This, together with Theorem 2.2.8 and (3.2.9) proves that the Chebyshev polynomials $\mathcal{T}_n$ are orthogonal with respect to the inner product

$$\langle p, q \rangle := \int_{-1}^{1} \frac{p(t)q(t)}{\sqrt{1-t^2}}dt$$

3.2 Chebyshev and Optimal Polynomials; One Interval

and that the associated polynomials (Chebyshev polynomials of the second kind) are given by $\mathcal{U}_n(t) = T'_n(t)/n$. Note, that in view of Example 3.1.2, the inner product is defined by the equilibrium distribution for the unit interval.

We summarize.

Theorem 3.2.2 *The following conditions are equivalent*

(a) $\mathcal{T}_n(t; I)$ is the Chebyshev polynomial on $I = [-1, 1]$;

(b) There exist a monic polynomial $\mathcal{U}_{n-1}$ of degree $n - 1$ and a constant L_n with

$$\mathcal{T}_n^2(t; I) + (1 - t^2)\mathcal{U}_{n-1}^2(t; I) = L_n^2;$$

(c) $\mathcal{T}_n(t; I)$ is the monic orthogonal polynomial on $[-1, 1]$ with respect to the weight function defined by the equilibrium distribution $dt(1 - t^2)^{-1/2}$;

(d) The optimal polynomial on $[-1, 1]$ with respect to the constraint point $\xi \notin [-1, 1]$ is

$$\mathcal{P}_n(t; I, \xi) = \frac{\mathcal{T}_n(t; I)}{\mathcal{T}_n(\xi; I)};$$

(e) $\mathcal{T}_n(t; I) = T_n(t)/2^{n-1}$, where T_n fulfills the three-term recurrence relation

$$T_0(t) = 1, \quad T_1(t) = t,$$
$$T_n(t) = 2t\, T_{n-1}(t) - T_{n-2}(t), \quad n \geq 2.$$

3.3 Chebyshev and Optimal Polynomials; Two Intervals

In this section we discuss the properties of Chebyshev polynomials and of optimal polynomials on the union of two disjoint intervals

$$\Omega = E = [a, b] \cup [c, d].$$

In fact, we will prove results similar to the one for the classical case as stated in Theorem 3.2.2.

Basic Observations

In accordance with the classical case it is convenient to map the set E onto the unit interval

$$l(E) = \hat{E} = [-1, \hat{b}] \cup [\hat{c}, 1],$$

where l denotes the linear transformation defined in (3.2.1). In the following we will discuss problem (3.1.3) and problem (3.1.4) for $\Omega = \hat{E}$.

Since Theorem 3.1.3 also holds for sets consisting of more than one component, the Chebyshev polynomial $T_n(t; \hat{E})$ for the union of two intervals is similarly characterized by an alternating set consisting of $n+1$ extremal points. Here, in contrast to the classical case, this condition does **not** imply that all boundary points $\{-1, \hat{b}, \hat{c}, 1\}$ are extremal points. Let us present an example for this situation.

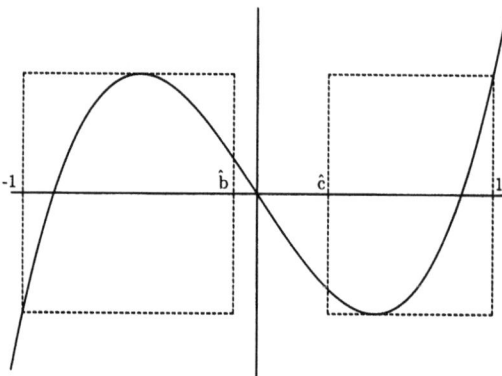

Figure 3.3.1 Chebyshev polynomial $T_n(t; \hat{E})$ of degree $n = 3$ with $n+1$ extremal points.

3.3 Chebyshev and Optimal Polynomials; Two Intervals

To discuss this example, let $t_2 \in (-1, \hat{b})$ and $t_3 \in (\hat{c}, 1)$ denote the interior extremal points of $\mathcal{T}_3(t; \hat{E})$. Then it is apparent from Figure 3.3.1, that $\mathcal{T}_3$ is the Chebyshev polynomial with respect to any set $[-1, \alpha] \cup [\beta, 1]$, where $t_2 \leq \alpha \leq \beta \leq t_3$. In other words, in this case, there is no one-to-one mapping between the Chebyshev polynomial and its "optimality region". This is an unfortunate situation. Moreover, observe that $\mathcal{T}_3$ has a zero ξ in the gap $[\hat{b}, \hat{c}]$. At least for this constraint point, the optimal polynomial $\mathcal{P}_3(t; \hat{E}, \xi)$ is **not** the scaled Chebyshev polynomials (cf. 3.2.3).

In the next sections we investigate Chebyshev polynomials which have none of these shortcomings.

One More Extremal Point

In this section we discuss Chebyshev and optimal polynomials of degree n with $n+2$ extremal points on $\hat{E}$. Recall that already $n+1$ alternation points uniquely characterize the best approximation. We remark that it is by no means obvious that there exist sets $\hat{E}$ where the corresponding Chebyshev (optimal) polynomial has $n+2$ extremal points. However, before we actually characterize those sets, let us investigate the question: what use can be made out of this property?

We start with an example.

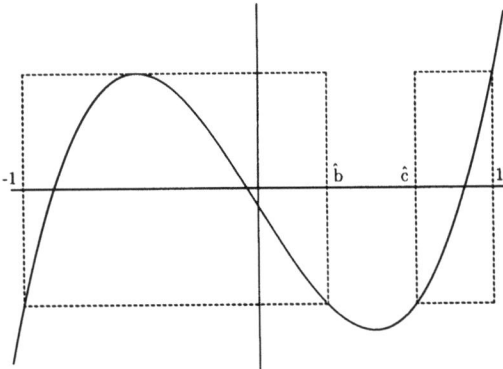

Figure 3.3.2 Chebyshev polynomial $\mathcal{T}_n(t; \hat{E})$ of degree $n = 3$ with $n+2$ extremal points.

The plot above illustrates the "general case". The proof of the next lemma is straightforward.

Lemma 3.3.3 Let $\hat{E} = [-1, \hat{b}] \cup [\hat{c}, 1]$ be given and let the Chebyshev polynomial $\mathcal{T}_n(t; \hat{E})$ have $n+2$ extremal points with $\|\mathcal{T}_n(t; \hat{E})\|_{\hat{E}} = L_n$.

(a) All four boundary points $-1, \hat{b}, \hat{c}, 1$ are extremal points of $\mathcal{T}_n(t; \hat{E})$.

(b) $\mathcal{T}_n(\hat{b}; \hat{E}) = \mathcal{T}_n(\hat{c}; \hat{E})$.

(c) $|\mathcal{T}_n(t; \hat{E})| > L_n$, for $t \in (\hat{b}, \hat{c})$.

(d) $\mathcal{T}'_n(t; \hat{E})$ has precisely one zero $t^* \in (\hat{b}, \hat{c})$.

Based on this lemma, we may now proceed as in classical case (compare (3.2.4) and (3.2.6)). For convenience, we again drop the dependence on the underlying region

$$\mathcal{T}_n(t) = \mathcal{T}_n(t; \hat{E}).$$

In view of Lemma 3.3.3(a) it follows that $\mathcal{T}_n$ attains its maximum at $n-2$ points $t_j \in (-1, \hat{b}) \cup (\hat{c}, 1)$, which are all zeros of $\mathcal{T}'_n$. With the help of Lemma 3.3.3(d), we derive the following identities (see, also Atlestam [7])

$$(3.3.1) \qquad (1 - t^2)(\hat{b} - t)(\hat{c} - t)(\mathcal{T}'_n(t))^2 = n^2(t^* - t)^2(L_n^2 - \mathcal{T}_n^2(t))$$

or, equivalently, with $U_{n-2}(t) := \mathcal{T}'_n(t)/(n(t - t^*))$,

$$(3.3.2) \qquad \mathcal{T}_n^2(t) + (1 - t^2)(\hat{b} - t)(\hat{c} - t)U_{n-2}^2(t) = L_n^2.$$

Following the lines of the classical case, we rewrite (3.3.1) as

$$\int \frac{\mathcal{T}'_n(t)}{\sqrt{L_n^2 - \mathcal{T}_n^2(t)}} dt = n \int \frac{|t^* - t|}{\sqrt{(1 - t^2)(\hat{b} - t)(\hat{c} - t)}} dt.$$

Note that the right hand side is an *elliptic integral*. One could now go ahead and evaluate this integral with the help of Jacobian elliptic functions

3.3 Chebyshev and Optimal Polynomials; Two Intervals

(see, e.g., Achieser [1, pp. 1198] and Byrd and Friedman [14]). We will not follow this approach. Instead we will first attack equation (3.3.2) to derive orthogonality relations for $\mathcal{T}_n$ and later on we will explicitly compute the Green's function with respect to the complement of $\hat{E}$ to come up with a result similar to the one stated in Cor. 3.2.1. This approach has the advantage that, as a by-product, we obtain a computable expression for the asymptotic convergence factor for $\hat{E}$ (cf. (3.1.6)).

Working from (3.3.2) we obtain for $|t| > 1$ that

$$\frac{U_{n-2}(t)}{\mathcal{T}_n(t)} = \frac{1}{\sqrt{(t^2-1)(t-\hat{b})(t-\hat{c})}} \sqrt{1 - \frac{L_n^2}{\mathcal{T}_n^2(t)}}$$

and, consequently (compare 3.2.9)

$$(3.3.3) \quad \frac{U_{n-2}(t)}{\mathcal{T}_n(t)} = \frac{1}{\sqrt{(t^2-1)(t-\hat{b})(t-\hat{c})}} + O\left(\frac{1}{t^{2n+2}}\right), \quad \text{for } |t| \to \infty.$$

Actually, we will make use of the next two relations, which follow directly from (3.3.3),

$$(3.3.4) \quad \frac{(t-\hat{b})U_{n-2}(t)}{\mathcal{T}_n(t)} = \frac{\sqrt{t-\hat{b}}}{\sqrt{(t^2-1)(t-\hat{c})}} + O\left(\frac{1}{t^{2n+1}}\right), \quad \text{for } |t| \to \infty,$$

and

$$(3.3.5) \quad \frac{(t-\hat{c})U_{n-2}(t)}{\mathcal{T}_n(t)} = \frac{\sqrt{t-\hat{c}}}{\sqrt{(t^2-1)(t-\hat{b})}} + O\left(\frac{1}{t^{2n+1}}\right), \quad \text{for } |t| \to \infty,$$

respectively. To apply Theorem 2.2.8 we need to further investigate the square roots of the righthand-sides of (3.3.4) and (3.3.5), respectively. The proof of the following lemma can be found in Peherstorfer [100, Lemma 2].

Lemma 3.3.4 Let $\hat{E} = [-1, \hat{b}] \cup [\hat{c}, 1]$ be given. Then we have for $t \notin [-1, 1]$

(a) $\dfrac{\sqrt{t-\hat{b}}}{\sqrt{(t^2-1)(t-\hat{c})}} = \displaystyle\int_{\hat{E}} \dfrac{w_{\hat{b}}(x)}{t-x} dx, \quad w_{\hat{b}}(x) := \dfrac{\sqrt{x-\hat{b}}}{\sqrt{(1-x^2)(x-\hat{c})}},$

(b) $\dfrac{\sqrt{t-\hat{c}}}{\sqrt{(t^2-1)(t-\hat{b})}} = \displaystyle\int_{\hat{E}} \dfrac{w_{\hat{c}}(x)}{t-x} dx, \quad w_{\hat{c}}(x) := \dfrac{\sqrt{x-\hat{c}}}{\sqrt{(1-x^2)(x-\hat{b})}}.$

This lemma together with Theorem 2.2.8 shows that Chebyshev polynomials $\mathcal{T}_n$ with $n+2$ extremal points are orthogonal with respect to the weight functions $w_{\hat{b}}$ and $w_{\hat{c}}$, respectively i.e.,

$$\int_{\hat{E}} \mathcal{T}_n(t) p(t) w_{\hat{b}}(t) dt = \int_{\hat{E}} \mathcal{T}_n(t) p(t) w_{\hat{c}}(t) dt = 0, \quad \text{for all} \quad p \in \Pi_{n-1}.$$

Observe, that we may recover the "one interval case" (cf. (3.2.10)) by noting that

$$w_{\hat{b}}(t) = w_{\hat{c}}(t) = \frac{1}{\sqrt{1-t^2}}, \quad \text{for} \quad \hat{b} = \hat{c}.$$

Moreover, since $\mathcal{T}_n$ is orthogonal with respect to any linear combination of $w_{\hat{b}}$ and $w_{\hat{c}}$, it is in particular orthogonal with respect to

$$w^*(t) := \frac{|t^* - t|}{\sqrt{(1-t^2)(t-\hat{b})(t-\hat{c})}},$$

where t^* is the relative extremum of $\mathcal{T}_n$ in the gap $(\hat{b}, \hat{c})$ (cf. Lemma 3.3.3(d)). This special case goes back to Achieser [2]. We mention that w^* is the equilibrium distribution for $\hat{E}$, as was shown by Geronimo and van Assche [63, Theorem 11] and by Peherstorfer [99, Lemma 2.2]).

Next, we state the "disjoint analogy" to Theorem 3.2.2.

3.3 Chebyshev and Optimal Polynomials; Two Intervals

Theorem 3.3.5 *Let $\hat{E} = [-1, \hat{b}] \cup [\hat{c}, 1]$ be given and let $w_{\hat{b}}$ and $w_{\hat{c}}$ be defined as in Lemma 3.3.4. Then, the following conditions are equivalent*

(a) $\mathcal{T}_n(t; \hat{E})$ *is the Chebyshev polynomial on $\hat{E}$ with $n+2$ extremal points on $\hat{E}$;*

(b) *There exist a monic polynomial $\mathcal{U}_{n-2}$ of degree $n - 2$ and a constant L_n with*

$$\mathcal{T}_n^2(t; \hat{E}) + (1 - t^2)(\hat{b} - t)(\hat{c} - t)\mathcal{U}_{n-2}^2(t; \hat{E}) = L_n^2;$$

(c) $\mathcal{T}_n(t; \hat{E})$ *is the monic orthogonal polynomial on $\hat{E}$ with respect to the weight function w, where w is given by $w_{\hat{b}}$, $w_{\hat{c}}$ or w^*;*

(d) *The optimal polynomial on $\hat{E}$ with respect to the constraint point $\xi \in (\hat{b}, \hat{c})$ is*

$$\mathcal{P}_n(t; \hat{E}, \xi) = \frac{\mathcal{T}_n(t; \hat{E})}{\mathcal{T}_n(\xi; \hat{E})}.$$

Proof. The equivalence of (a), (b), and (c) follows from the considerations above (see also Peherstorfer [100, Theorem 4]).

It remains to show that (a) is equivalent to (d). We follow Fischer [37, Theorem 4.3]. In view of Cor. 3.1.4(a), the optimal polynomial $\mathcal{P}(t; \hat{E}, \xi)$ is uniquely characterized by the existence of $n + 1$ extremal points $t_1 < t_2 < \cdots < t_{n+1} \in \hat{E}$ with

$$\mathcal{P}_n(t_j; \hat{E}, \xi) = (-1)^{\tau+1}\mathcal{P}_n(t_{j+1}; \hat{E}, \xi), \quad \text{where} \quad \tau = \begin{cases} 1 & \text{if } \xi \in [t_j, t_{j+1}] \\ 0 & \text{otherwise.} \end{cases}$$

The Chebyshev polynomial $\mathcal{T}_n(t); \hat{E})$ is essentially characterized by the same requirements, except that $\tau = 0$ for all j. First, let us assume that $\mathcal{T}_n(z; E)$ has $n+2$ extremal points t_j (compare Figure 3.3.2). Cor. 3.1.4(b), together with Lemma 3.3.3(b), immediately yields

$$\mathcal{T}_n(t_j; \hat{E}) = (-1)^{\tau+1}\mathcal{T}_n(t_{j+1}; \hat{E}), \quad \text{where} \quad \tau = \begin{cases} 1 & \text{if } t_j = \hat{b} \text{ and } t_{j+1} = \hat{c} \\ 0 & \text{otherwise,} \end{cases}$$

which is all we need.

The proof for the other direction goes by contradiction. Suppose, to the contrary, that $\mathcal{T}_n(z; \hat{E})$ has only $n+1$ extremal points $t_1 < t_2 < \cdots < t_{n+1}$ (compare Figure 3.3.1). Then it follows that $\mathcal{T}_n$ alternates at each of this points, i.e., $\mathcal{T}_n(t_j; \hat{E}) = -\mathcal{T}_n(t_{j+1}; \hat{E})$, $j = 1, 2, \ldots, n+1$. Hence, $\mathcal{P}_n(t; \hat{E}, \xi)$ is not a multiple of the Chebyshev polynomial. □

We remark that part (d) includes the result of Lebedev [88] for the symmetric case $\hat{b} = -\hat{c}$.

We stress that Theorem 3.3.5(c) does **not** imply that $\mathcal{T}_k$, $k \neq n$, is orthogonal with respect to $w_{\hat{b}}$ or $w_{\hat{c}}$. This statement is in general not true. On the other hand, however, the next corollary shows that the situation is not as bad. The proof reduces to a careful counting of extremal points.

Corollary 3.3.6 Let $\hat{E} = [-1, \hat{b}] \cup [\hat{c}, 1]$ be given and let $\mathcal{T}_n(t; \hat{E})$ have $n+2$ extremal points on $\hat{E}$ with $L_n = \|\mathcal{T}_n(t; \hat{E})\|_{\hat{E}}$. Furthermore, let $\mathcal{T}_n(t; I)$ denote the Chebyshev polynomial on the unit interval. Then, we have

(a) The Chebyshev polynomial of degree $n \cdot l$, $l \in \mathbb{N}$, is given by

$$\mathcal{T}_{n \cdot l}(t; \hat{E}) = \mathcal{T}_l\left(\mathcal{T}_n(t; \hat{E})/L_n; I\right),$$

where $\mathcal{T}_{n \cdot l}(t; \hat{E})$ has $nl + 2$ extremal points on $\hat{E}$.

(b) The optimal polynomial of degree $n \cdot l + 1$, $l \in \mathbb{N}$, is "defective"

$$\mathcal{P}_{n \cdot l+1}(t; \hat{E}, \xi) = \mathcal{P}_{n \cdot l}(t; \hat{E}, \xi) = \frac{\mathcal{T}_{n \cdot l}(t; \hat{E})}{\mathcal{T}_{n \cdot l}(\xi; \hat{E})}, \quad \text{for } \xi \notin \hat{E}.$$

3.3 Chebyshev and Optimal Polynomials; Two Intervals

Clearly, there exist orthogonal polynomials, of any degree, with respect to $w_{\hat{b}}$ and $w_{\hat{c}}$. The next theorem provides explicit expressions for their three-term recurrence coefficients. The proof can be found in Peherstorfer [98, Satz 2, Korollar 1].

Theorem 3.3.7 Let $\hat{E} = [-1, \hat{b}] \cup [\hat{c}, 1]$ be given and let $w_{\hat{b}}$ and $w_{\hat{c}}$ be defined as in Lemma 3.3.4.

(a) Let $\sigma := (\hat{c} - \hat{b})/2$ and $\tau := (\hat{c} + \hat{b})/2$. Then, the monic orthogonal polynomials ψ_n^b with respect to $w_{\hat{b}}$ on $\hat{E}$ are given by

$$\psi_0^b(t) = 1, \quad \psi_1^b(t) = t - \sigma,$$
$$\psi_n^b(t) = (t - \alpha_n)\psi_{n-1}^b(t) - \beta_n \psi_{n-2}^b(t), \quad n > 1,$$

where

$$\beta_n = \begin{cases} (1 - \hat{b}^2 + \tau^2)/2 & \text{for } n = 2 \\ \alpha_2(\tau - \alpha_2) + (1 - \beta_2 + \sigma^2)/2 & \text{for } n = 3 \\ \alpha_{n-1}(\tau - \alpha_{n-1}) - \beta_{n-1} + (1 + \sigma^2)/2 & \text{for } n > 3, \end{cases}$$

and

$$\alpha_n = \begin{cases} -(\hat{c} + \hat{b}\sigma^2)/(2\beta_2) + \tau & \text{for } n = 2 \\ \beta_2(\alpha_2 - \hat{b})/(2\beta_3) + \tau - \alpha_2 & \text{for } n = 3 \\ \beta_{n-1}(\alpha_{n-1} + \alpha_{n-2} - \tau)/\beta_n + \tau - \alpha_n & \text{for } n > 3. \end{cases}$$

(b) The three-term recurrence coefficients of the monic orthogonal polynomials ψ_n^c with respect to $w_{\hat{c}}$ on $\hat{E}$ are obtained from the one of ψ_n^b by interchanging the role of $\hat{b}$ and $\hat{c}$.

Next, we state a corollary to the proof of Theorem 3.3.5. It says that the number of extremal points plays no role for the case of constraint points to the left or to the right of $\hat{E}$.

> **Corollary 3.3.8** Let $\hat{E} = [-1, \hat{b}] \cup [\hat{c}, 1]$ and let $\xi \notin [-1, 1]$ be given. Then
> $$\mathcal{P}_n(t; \hat{E}, \xi) = \frac{\mathcal{T}_n(t; \hat{E})}{\mathcal{T}_n(\xi; \hat{E})}.$$

So, we are in "good shape" as long as the Chebyshev polynomial $\mathcal{T}_n(t; \hat{E})$ has $n + 2$ extremal points. It remains to characterize such sets $\hat{E}$. This will be topic of the next sections. Here, the key tool will be the complex Green's function $\mathcal{G}(t; \hat{E}, 0)$ with respect to $\hat{E}^c$. It will turn out, that the Chebyshev polynomial admits a representation in terms of the complex Green's function, if, and only if $\mathcal{T}_n(t; \hat{E})$ has $n + 2$ extremal points (compare Cor. 3.2.1).

To compute $\mathcal{G}(t; \hat{E}, 0)$ we will, in accordance with Example 3.1.1(b), first determine the conformal mapping

$$\phi_{\hat{E}} : A(\delta) \to \hat{E}^c$$

from a suitable annulus, the canonical region of doubly-connected regions,

(3.3.6) $$A(\delta) := \{z \in \mathbb{C} : \delta < |z| < 1\}, \quad 0 < \delta < 1,$$

onto $\hat{E}^c$. Then we will construct a complex Green's function for $A(\delta)$ and finally derive $\mathcal{G}(t; \hat{E}, 0)$ via conformal transplantation.

All these derivations will involve *elliptic functions*. For those readers who are not so familiar with these functions, we will start by collecting some basic facts from this subject.

Elliptic Functions

An introduction to the theory of elliptic functions can be found in Whittaker and Watson [133]. Here we provide some basic facts and collect some auxiliary results.

A meromorphic function f is said to be *elliptic* if it is doubly-periodic

$$f(z + \omega_1) = f(z), \ f(z + \omega_2) = f(z), \ \text{where } \operatorname{Im}(\omega_1/\omega_2) > 0.$$

3.3 Chebyshev and Optimal Polynomials; Two Intervals 111

The set of periods $\sigma w_1 + \tau w_2$, $\sigma, \tau = 0, \pm 1, \pm 2, \ldots$, (we assume that all the periods of f are of this form, i.e., w_1 and w_2 are *primitive periods*) forms a lattice in the plane and determines a set of congruent parallelograms, known as *period parallelograms of f*. The sum of the multiplicities of the poles of an elliptic function in one period parallelogram is called the *order* of the function. We collect some fundamental properties of elliptic functions, which we shall have occasion to use, in the following lemma. It turns out that elliptic functions share quite a few properties with "ordinary rational functions".

Lemma 3.3.9

(a) *The system of elliptic functions having the same set of periods forms a field.*

(b) *If two elliptic functions have the same periods, poles, and zeros, then their ratio is a constant.*

(c) *An elliptic function of order $n \geq 2$ assumes every value n times (each counted according to its multiplicities).*

Let us now derive and discuss some "prominent" elliptic functions. Instead of considering the elliptic integral

(3.3.7)
$$z = F(w; k) = \int_0^w \frac{dt}{\sqrt{(1-t^2)(1-k^2 t^2)}}$$
$$= \int_0^v \frac{dx}{\sqrt{1 - k^2 \sin^2 x}} = \hat{F}(v; k),$$

Abel and Jacobi investigated the inversion of this integral. One may define

$$\operatorname{sn}(z; k) := F^{-1}(z; k) = w = \sin(v) \quad \text{and} \quad \operatorname{am}(z; k) := \hat{F}^{-1}(z; k) = v.$$

The function $\operatorname{sn}(z; k) = \sin(\operatorname{am}(z; k))$ is known as *elliptic sine* (*sinus amplitudinis*), whereas $\operatorname{am}(z; k)$ is called *amplitude function of z*. The number k, $0 < k < 1$, is known as *modulus*.

To set the approach above into perspective, we remark that the ordinary sine is just the inverse $w = \sin(z)$ of the integral

$$z = \int_0^w \frac{dt}{\sqrt{1-t^2}}, \quad w \leq 1.$$

Closely connected with sn is the *elliptic cosine* (*cosinus amplitudinis*) and the *delta amplitudinis*

(3.3.8)
$$\operatorname{cn}(z;k) := \cos(\operatorname{am}(z;k)) = \sqrt{1 - \operatorname{sn}(z;k)^2},$$
$$\operatorname{dn}(z;k) := \Delta(\operatorname{am}(z;k)) = \sqrt{1 - k^2 \operatorname{sn}(z;k)^2}.$$

The functions sn, cn, and dn are known as *Jacobian elliptic functions*.

Furthermore, of great importance is the *complete elliptic integral of the first kind with respect to the modulus k*, defined by

(3.3.9) $$K(k) := \operatorname{sn}^{-1}(1;k) = \int_0^1 \frac{dt}{\sqrt{(1-t^2)(1-k^2 t^2)}},$$

and the *complete elliptic integral of the first kind with respect to the complementary modulus k'*, defined by

$$K'(k) := K(k'), \quad k' := \sqrt{1 - k^2}.$$

Let us introduce a notational convention. In most cases we will drop the dependence on the modulus k. Only when it is helpful, we will indicate the modulus.

As for the trigonometric functions there are numerous relations between the Jacobian elliptic functions (cf. Whittaker and Watson [133, pp. 491] for a collection). We will have occasion to use

(3.3.10) $$\operatorname{sn}(u + K) = \frac{\operatorname{cn}(u)}{\operatorname{dn}(u)}.$$

For real arguments, the elliptic sine and elliptic cosine mimic the ordinary sine and cosine, respectively. In fact, we have

$$\sin(t) = \operatorname{sn}(t;0) \quad \text{and} \quad \cos(t) = \operatorname{cn}(t;0).$$

3.3 Chebyshev and Optimal Polynomials; Two Intervals

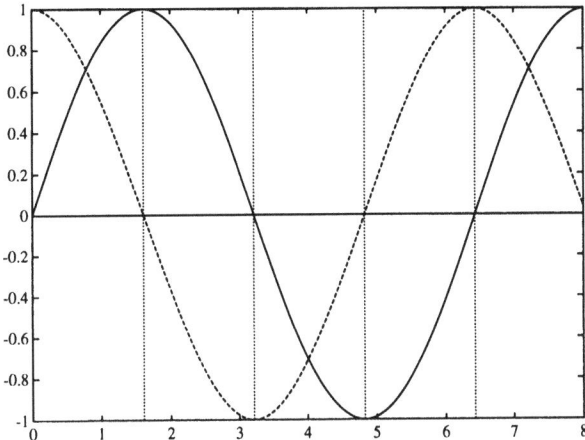

Figure 3.3.10 Solid curve: elliptic sine for $k = 0.3$; dashed curve: elliptic cosine for $k = 0.3$; dotted lines: $t = j * K(0.3)$ where $K(0.3) = 1.608\ldots$

In the next lemma, we list some of the properties of sn and cn (cf. Whittaker and Watson [133, pp. 491]).

Lemma 3.3.11

(a) The elliptic sine $\text{sn}(z)$ is an odd elliptic function of order 2 with the primitive periods $4K$ and $2iK'$. Moreover, it holds

$$\text{sn}(z + 2K) = -\text{sn}(z).$$

Finally, sn has simple zeros at $z \equiv 0 (\text{mod } 2K, 2iK')$ and simple poles at $z \equiv iK' (\text{mod } 2K, 2iK')$.

(b) The elliptic cosine $\text{cn}(z)$ is an even elliptic function of order 2 with the primitive periods $4K$ and $2(K + iK')$. Moreover, it holds

$$\text{cn}(z + 2K) = \text{cn}(z + 2iK) = -\text{cn}(z).$$

Finally, cn has simple zeros at $z \equiv K (\text{mod } 2K, 2iK')$ and simple poles at $z \equiv iK' (\text{mod } 2K, 2iK')$.

One method for the computation of sn and cn is provided by the relations

$$\operatorname{sn}(z) = \frac{1}{\sqrt{k}} \frac{H(z)}{\theta(z)}, \quad \operatorname{cn}(z) = \sqrt{\frac{k'}{k}} \frac{H(z+K)}{\theta(z)}.$$

Here H and θ are defined by

(3.3.11)
$$H(z) := \vartheta_1(\frac{z}{2K}\pi; q) = 2q^{1/4} \sum_{n=0}^{\infty} (-1)^n q^{n(n+1)} \sin(\frac{(2n+1)z}{2K}\pi),$$

$$\theta(z) := \vartheta_4(\frac{z}{2K}\pi; q) = 1 + 2\sum_{n=1}^{\infty} (-1)^n q^{n^2} \cos(\frac{2nz}{2K}\pi),$$

The functions H and θ are known as *Theta functions* with respect to the nome $q := \exp(-\pi K'/K)$, where we have used Jacobi's earlier notation. We collect some properties of H and θ in the following lemma (cf. Whittaker and Watson [133, pp. 462]).

Lemma 3.3.12

(a) $H(z)$ is an odd function which is analytic at every point of the complex plane. It is quasi doubly-periodic

$$H(z + 2K) = -H(z)$$
$$H(z + 2iK') = -e^{-i\pi z/K} q^{-1} H(z)$$
$$H(z + iK') = ie^{-i\pi z/(2K)} q^{-1/4} \theta(z),$$

and satisfies the symmetry relation $H(\bar{z}) = \overline{H(z)}$. Moreover, H has simple zeros at $z \equiv 0 \pmod{2K, 2iK'}$.

(b) $\theta(z)$ is an even function which is analytic at every point of the complex plane. it is quasi doubly-periodic

$$\theta(z + 2K) = \theta(z),$$
$$\theta(z + 2iK') = -e^{-i\pi z/K} q^{-1} \theta(z),$$

and satisfies the symmetry relation $\theta(\bar{z}) = \overline{\theta(z)}$. θ has simple zeros at $z \equiv iK' \pmod{2K, 2iK'}$.

3.3 Chebyshev and Optimal Polynomials; Two Intervals

Having introduced the elliptic functions we are ready to discuss the conformal mapping $\phi_{\hat{E}}$ form the annulus $A(\delta)$ onto $\hat{E}$ (cf. (3.3.6)).

A Conformal Mapping

The desired conformal mapping $\phi_{\hat{E}} = \chi \circ \varphi$ will be composed out of two mappings

$$A(\delta) \xrightarrow{\chi} R \xrightarrow{\varphi} \hat{E},$$

where R denotes a certain rectangle.

Let us start by constructing the mapping φ and the rectangle R. It is advantageous to parameterize the interior boundary points $\hat{b}$ and $\hat{c}$ of $\hat{E}$. To this end let

(3.3.12) $$k^2 := \frac{2(\hat{c} - \hat{b})}{(1 - \hat{b})(1 + \hat{c})} \quad (< 1)$$

and take k as the modulus of the elliptic functions. Next, determine the number ρ, $-K < \rho < 0$, from the identity (cf. Figure 3.3.10)

(3.3.13) $$\hat{b} = 1 - 2\mathrm{sn}^2(\rho),$$

It follows from (3.3.12) and (3.3.13) and (3.3.10) that

(3.3.14) $$\hat{c} = 2\frac{\mathrm{cn}^2(\rho)}{\mathrm{dn}^2(\rho)} - 1 = 2\mathrm{sn}^2(\rho + K) - 1.$$

For convenience we will use the notation

(3.3.15) $$\{\hat{E}, \hat{b}, \hat{c}, k, \rho\}$$

for the set $\hat{E} = [-1, \hat{b}] \cup [\hat{c}, 1]$ with the "parameter" k and ρ as computed in (3.3.12) and (3.3.13), respectively.

We remark that the other way round any choice of $k \in (0, 1)$ and $\rho \in (-K, 0)$ uniquely determines, by (3.3.13) and (3.3.14), a set of points $\hat{b}, \hat{c} \in [-1, 1]$. Also, we note for future reference that ρ in (3.3.13) is given by

(3.3.16) $$\rho = -\mathrm{sn}^{-1}(w_\rho), \quad \text{where} \quad w_\rho = \sqrt{\frac{1 - \hat{b}}{2}}.$$

Now consider

(3.3.17) $$z = \varphi(u; \rho) := \frac{\operatorname{sn}^2(u)\operatorname{cn}^2(\rho) + \operatorname{cn}^2(u)\operatorname{sn}^2(\rho)}{\operatorname{sn}^2(u) - \operatorname{sn}^2(\rho)}.$$

It is not hard to check (cf. Achieser [3, pp. 138] and Kober [86, pp.191]) that φ describes a one-to-one mapping of the (partly open) rectangle

(3.3.18) $$R = \{ u \in \mathbb{C} : -K < \operatorname{Re} u < 0, \ -iK' < \operatorname{Im} u \leq iK' \}$$

onto $\hat{E}^c$.

We collect some properties of φ in the next lemma.

Lemma 3.3.13 *Let $\{\hat{E}, \hat{b}, \hat{c}, k, \rho\}$ (cf. (3.3.15)) be given.*

(a) $\varphi(u; \rho)$ is an even elliptic function of order 2 with the primitive periods $2K$ and $2iK'$. Moreover, φ has simple poles at $\pm \rho$.

(b) $\varphi(u; \rho)$ may be rewritten as

$$\varphi(u; \rho) = \hat{b} + \frac{1 - \hat{b}^2}{2\operatorname{sn}^2(u) + \hat{b} - 1}.$$

(c) The set $\hat{E}$ is the image of the vertical sides of R

$$\varphi([0, iK']; \rho) = [-1, \hat{b}] \quad \text{and} \quad \varphi([-K + iK', -K]; \rho) = [\hat{c}, 1].$$

and the gap $[\hat{b}, \hat{c}]$ is the image of the (upper) horizontal side

$$\varphi([-K + iK', iK']; \rho) = [\hat{b}, \hat{c}].$$

(d) Let $z \in (\hat{b}, \hat{c})$ be given. Then $\varphi(u + iK'; \rho) = z$ for

$$u = -\operatorname{sn}^{-1}(w_u), \quad \text{where} \quad w_u = -\frac{1}{k \operatorname{sn}(\rho)}\sqrt{\frac{z - \hat{b}}{z + 1}}.$$

3.3 Chebyshev and Optimal Polynomials; Two Intervals

Proof. The proof of (a) and (b) is a straightforward application of Lemma 3.3.11.

To prove (c) we need quite a few properties of the Jacobian elliptic functions. They all can be found in Whittaker and Watson [133, pp. 491]. To start with, recall that $k' = \sqrt{1-k^2}$ denotes the complementary modulus. Then, it holds

$$\text{sn}(iu; k) = i\frac{\text{sn}(u; k')}{\text{cn}(u; k')} \quad \text{and} \quad \text{cn}(iu; k) = \frac{1}{\text{cn}(u; k')},$$

which implies that $\varphi(u; \rho)$ attains real values on the segment $[0, iK']$. Furthermore, we have

$$\text{sn}(0; k) = 0, \quad \text{cn}(0; k) = 1, \quad \text{and} \quad \text{sn}(iK'; k) = \infty,$$

which shows, with the help of (b), that

$$\varphi(0; \rho) = -1 \quad \text{and} \quad \varphi(iK'; \rho) = \hat{b}.$$

Next, we apply the identities

(3.3.19)
$$\text{sn}(-tK + iK'; k) = -\frac{1}{k\,\text{sn}(tK; k)}$$

$$\text{cn}(-tK + iK'; k) = \frac{i\,\text{dn}(tK; k)}{k\,\text{sn}(tK; k)}$$

to prove that $\varphi(u; \rho)$ attains real values on the segment $[-K + iK', iK']$. In particular, we have by (3.3.19), (3.3.14), $\text{sn}(K; k) = 1$, (3.3.8), and $\text{dn}(K; k) = k'$ that

$$\varphi(-K + iK'; \rho) = \frac{\text{cn}^2(\rho; k) - (1-k^2)\text{sn}^2(\rho; k)}{1 - k^2\text{sn}^2(\rho; k)} = \frac{2\text{cn}^2(\rho; k)}{\text{dn}^2(\rho; k)} - 1 = \hat{c}.$$

Moreover, by $\text{sn}(-K; k) = -1$ and (b) we have

$$\varphi(-K; \rho) = 1.$$

With

$$\text{sn}(-K + itK'; k) = -\frac{1}{\text{dn}(tK'; k')} \quad \text{and} \quad \text{cn}(-K + itK'; k) = ik'\text{sn}(tK'; k')$$

it follows that $\varphi(u;\rho)$ attains real values on the segment $[-K, -K+iK']$.
Finally, observe that $\varphi(u;\rho)$ attains each value precisely once on the segments under consideration.

Statement (d) is a direct consequence of (3.3.19). □

The next figure illustrates the correspondence between points in the u-plane and points in the z-plane.

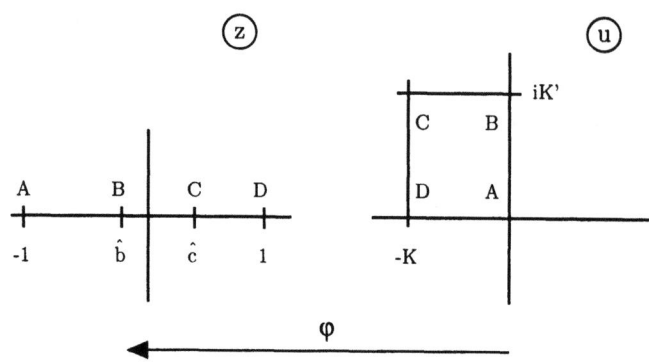

Figure 3.3.14 Corresponding points of the mapping φ.

Next, consider the mapping

$$u = \chi(v) := \frac{K'}{\pi} \log(v)$$

where the branch of the logarithm is chosen such that $\log(v) = \log|v| + i \arg v$, with $-\pi < \arg v \leq \pi$. The function χ maps the open annulus $A(\delta)$ (cf. (3.3.6)) bounded by the circles $|v| = 1$ and $|v| = \delta = \exp(-\pi K/K')$ and cut along the segment in the negative half of the real axis onto the rectangle R (cf. (3.3.18)). To remove this cut, one has to identify the upper side and lower side of R in the u-plane. Here, Lemma 3.3.12(a) implies

$$\varphi(u - iK'; \rho) = \varphi(u + iK'; \rho)$$

which is just the desired property. Altogether we have shown that

(3.3.20) $$z = \phi_{\hat{E}}(v) = \varphi(\chi(v); \rho) = \hat{b} + \frac{1 - \hat{b}^2}{2\mathrm{sn}^2\left(K' \log(v)/\pi\right) + \hat{b} - 1}$$

3.3 Chebyshev and Optimal Polynomials; Two Intervals

is the conformal mapping from $A(\delta)$ onto $\hat{E}$. Notice that the point ∞ in the z-plane corresponds to

(3.3.21) $$e^{\pi \rho/K'} = \chi^{-1}(\rho)$$

in the v-plane.

Green's Function for the Union of Two Disjoint Intervals

Using standard techniques from complex analysis (cf. Courant, Hilbert [23, pp. 312], Henrici [80, pp. 259] it is not hard to devise a formula for a complex Green's function for $A(\delta)$ (cf. (3.3.6)). Here we follow the construction of Achieser [1, pp. 1174].

Consider the function

(3.3.22) $$G(u;\rho) := \frac{H(u-\rho)}{H(u+\rho)}.$$

defined in the u-plane by the Theta function H (cf. (3.3.11)). With the help of Lemma 3.3.12(a) we can study properties of G.

Lemma 3.3.15 *Let $k \in (0,1)$ and $\rho \in (-K,0)$ be given. Then $G(u;\rho)$ is an analytic function in R (cf. (3.3.18)) with precisely one (simple) zero $u_0 = \rho$ in R. Moreover, G has modulus one on the vertical sides of the rectangle*

$$|G(-K+tiK';\rho)| = |G(tiK';\rho)| = 1, \ t \in [-1,1],$$

is quasi doubly-periodic

$$G(u+2K;\rho) = G(u;\rho), \ G(u+2iK';\rho) = e^{2i\pi\rho/K} G(u;\rho),$$

and satisfies

$$G(-u;\rho) = \frac{1}{G(u;\rho)}.$$

Proof. The periodic properties follow directly from Lemma 3.3.12(a). To verify the remaining part, observe that

$$\overline{G(-K+tiK';\rho)} = G(-K-tiK';\rho) = G(K-tiK';\rho) = \frac{1}{G(-K+tiK';\rho)},$$

and

$$\overline{G(tiK';\rho)} = G(-tiK';\rho) = \frac{1}{G(tiK';\rho)},$$

which concludes the proof. □

With the help of Lemma 3.3.12 and of Lemma 3.3.15 it is now straightforward to verify that

$$(3.3.23) \qquad \mathcal{G}(v; A(\delta), v_0) := G(\chi(v); \chi(e^{\pi\rho/K'})) = \frac{H(K'\log(v)/\pi - \rho)}{H(K'\log(v)/\pi + \rho)}$$

is a complex Green's function for the annulus $A(\delta)$ with the simple zero $v_0 = \exp(\pi\rho/K')$ in $A(\delta)$ (cf. (3.3.23)). Hence, we conclude by (3.1.7) that

$$(3.3.24) \qquad g(v; A(\delta), v_0) = -\log(|G(\chi(v); \chi(v_0))|)$$

is the Green's function for $A(\delta)$ with pole at v_0.

Altogether, we arrive at

Theorem 3.3.16 *Let $\{\hat{E}, \hat{b}, \hat{c}, k, \rho\}$ (cf. (3.3.15)) be given. Furthermore, let $\varphi(u;\rho)$ and $G(u;\rho)$ be defined as in (3.3.17) and (3.3.22), respectively. Then, the Green's function for $\hat{E}^c$ with pole at infinity is given by*

$$g(z; \hat{E}^c, \infty) = -\log(|G(u;\rho)|), \quad z = \varphi(u;\rho).$$

Proof. We apply the conformal mapping $\phi_{\hat{E}}$ (cf. (3.3.20)), the representation (3.3.24), (3.3.21), and the method of conformal transplantation

3.3 Chebyshev and Optimal Polynomials; Two Intervals

(compare Example 3.1.1(b)) to obtain

$$\begin{aligned}g(z;\hat{E}^c,\infty) &= g(\phi_{\hat{E}}^{-1}(z);\phi_{\hat{E}}^{-1}(\hat{E}^c),\phi_{\hat{E}}^{-1}(\infty))\\ &= g(\phi_{\hat{E}}^{-1}(z);A(\delta),v_0)\\ &= -\log(|G(\chi(\phi_{\hat{E}}^{-1}(z));\chi(v_0))|)\\ &= -\log(|G(u;\rho)|),\end{aligned}$$

where $z = \varphi(u;\rho)$. $\square$

A little remark is appropriate here. It might appear somewhat puzzling that we represented the Green's function in terms of a function defined in the u-plane and not in terms of a function defined in the v-plane. However, due to the rich properties of elliptic functions it will turn out to be much more convenient to work in the u-plane. First examples of this statement are the proofs of the next section.

The Achieser Representation of the Chebyshev Polynomials

We are now in position to state the announced representation of $\mathcal{T}_n(t;\hat{E})$ in terms of a complex Green's function. Recall from Cor. 3.2.1 that the Chebyshev polynomial $\mathcal{T}_n(t;I)$ for the unit interval I can always be expressed in terms of an associated complex Green's function

$$\mathcal{T}_n(t;I) = \frac{1}{2^n}\left(G(v;D,0)^n + \frac{1}{G(v;D,0)^n}\right), \quad t = \phi_I(v).$$

Let us carry over this "construction principle" to the set $\hat{E}$. More precisely, consider the function(s)

$$(3.3.25) \quad q_n(z;\rho) := Q_n(u;\rho) := \frac{1}{2}\left(G(u;\rho)^n + \frac{1}{G(u;\rho)^n}\right), \quad z = \varphi(u;\rho),$$

where G and φ are given in (3.3.22) and (3.3.17), respectively. We stress that the same construction could have been done in the v-plane.

At a first glance it seems rather unlikely that q_n is a polynomial in z. However, the following theorem holds.

> **Theorem 3.3.17** Let $k \in (0,1)$ and $\rho \in (-K, 0)$ be given and let $q_n(z; \rho)$ be defined by (3.3.25). Then, $q_n(z; \rho)$ is a polynomial in z if, and only if
> $$\rho = \rho_n := -\frac{m}{n}K, \quad \text{where} \quad m \in \{1, 2, \ldots, n-1\}.$$

Proof. In view of the representation (3.3.25), we have to show that there exist coefficients a_j with

$$(3.3.26) \qquad Q_n(u; \rho) = \sum_{j=0}^{n} a_j \varphi^j(u; \rho).$$

By Lemma 3.3.9 this is possible only if $Q_n(u; \rho)$ is an elliptic function with periods $2K$ and $2iK'$. It follows from Lemma 3.3.12(a) and Lemma 3.3.15 that

$$Q_n(u + 2K; \rho) = Q_n(u; \rho)$$
$$Q_n(u + 2iK'; \rho) = \left(e^{2in\pi\rho/K} G(u; \rho)^n + e^{-2in\pi\rho/K} G(u; \rho)^{-n}\right)/2$$
$$Q_n(-u; \rho) = Q_n(u; \rho).$$

Hence, $Q_n(u; \rho)$, $\rho \in (-K, 0)$, is an elliptic function with periods $2K$ and $2iK'$ if
$$\rho = -\frac{m}{n}K, \quad \text{where} \quad m \in \{1, 2, \ldots, n-1\},$$
which proves one direction.

Now, assume that $\rho = -mK/n$, i.e., Q_n is an even elliptic function with periods $2K$ and $2iK'$. This part of the proof is based on the fact that any even elliptic function with periods $2K$ and $2iK'$ can be represented as a rational function in sn^2 (cf. Achieser [3, §26]). So, $Q_n(u; \rho)$ admits such a representation. In view of Lemma 3.3.9 the associated rational function has to match the zeros and poles of Q_n. From Lemma 3.3.15 we know that $Q_n(u; \rho)$ has only the poles $\pm \rho$ (of order n) in its period parallelogram. Finally, recall from Lemma 3.3.12(a) that $\varphi(u, \rho)$ is an even elliptic function

3.3 Chebyshev and Optimal Polynomials; Two Intervals

with periods $2K$ and $2iK'$ and the simple poles $\pm\rho$, which, altogether, leaves only the representation (3.3.26) for Q_n. □

Next, we formulate the main result of this section. It says that if q_n is a polynomial, then it is, up to a constant, the Chebyshev polynomial (cf. 3.1.4) for the set $\hat{E}$ defined by k and ρ_n. The next theorem is due to Achieser [1, pp. 1190]

Theorem 3.3.18 Let $k \in (0,1)$ and let $\rho_n = -mK/n$ with $m \in \{1, 2, \ldots, n-1\}$ be given. Moreover, let $q_n(z;\rho)$ be defined by (3.3.25). Finally, let $\hat{b} = \hat{b}(k, \rho_n)$ and $\hat{c} = \hat{c}(k, \rho_n)$ be computed by (3.3.13) and (3.3.14), respectively. Then

$$\mathcal{T}_n(z; \hat{E}) = L_n q_n(z; \rho_n)$$

is the Chebyshev polynomial with respect to $\hat{E} = [-1, \hat{b}] \cup [\hat{c}, 1]$. The minimal deviation is given by

$$\|\mathcal{T}_n(z; \hat{E})\|_{\hat{E}} = L_n, \quad \text{where} \quad L_n = \frac{1}{2^{n-1}} \left(\frac{\theta(0)\theta(K)}{\theta(\rho_n)\theta(\rho_n + K)} \right)^{2n}.$$

Moreover, $\mathcal{T}_n(z; \hat{E})$ has $n+2$ extremal points on $\hat{E}$, namely $n-m+1$ alternation points on $[-1, \hat{b}]$ and $m+1$ alternation points on $[\hat{c}, 1]$.

Proof. Recall from Lemma 3.3.13(c) that

$$\varphi([0, iK']; \rho_n) = [-1, \hat{b}] \quad \text{and} \quad \varphi([-K + iK', -K]; \rho_n) = [\hat{c}, 1].$$

Hence, in view of (3.3.25) we have to investigate the behavior of

$$Q_n(itK'; \rho_n) \quad \text{and} \quad Q_n(-K + itK'; \rho_n), \quad \text{for} \quad t \in [0, 1].$$

Let us start with the segment $[0, iK']$. From Lemma 3.3.15 we know that the extremal points of Q_n are the points $u \in [0, iK']$ with $|Q_n(u; \rho_n)| = 1$. These are just the points where

$$\arg(G(u; \rho_n)) = l\frac{\pi}{n}, \quad l \in \mathbb{N}.$$

It remains to trace the change in the argument of $G(u;\rho)$ when u runs through the segment of the imaginary axis from $u = 0$ to $u = iK'$. To this end let us assume that
$$\arg\left(G(0;\rho_n)\right) = 0.$$
Furthermore we deduce from Lemma 3.3.12 that
$$G(iK';\rho_n) = \frac{H(iK' + mK/n)}{H(iK' - mK/n)} = \frac{e^{-i\pi m/(2n)}}{e^{i\pi m/(2n)}} \frac{\theta(mK/n)}{\theta(-mK/n)} = e^{-i\pi m/n}.$$

Consequently, $Q_n(u;\rho_n)$ takes its maximum on $[0, iK']$ with alternating sign at least $n - m + 1$ times. With the same technique, one may show that $Q_n(u;\rho_n)$ takes its maximum on $[-K, -K + iK']$ with alternating sign at least $m + 1$ times. Now, the statement follows from the fact that T_n can have at most $n + 2$ extremal points on $\hat{E}$.

For the derivation of the minimal deviation we refer to Achieser [1]. □

The theorem above says that the Chebyshev polynomial for a given set is explicitly known, whenever the ratio ρ/K is rational. Since all participating functions are continuous, it is not hard to see that for any given set $\hat{E} = [-1, \hat{b}] \cup [\hat{c}, 1]$ and tolerance $\varepsilon > 0$ there exist, for sufficiently large n, numbers $\hat{b}_n$ and $\hat{c}_n$ with $\max\{|\hat{b} - \hat{b}_n|, |\hat{c} - \hat{c}_n|\} < \varepsilon$, such that the Chebyshev polynomial for the set $\hat{E}_n = [-1, \hat{b}_n] \cup [\hat{c}_n, 1]$ has $n + 2$ extremal points. This statement can be made more precise. The following theorem is due to Peherstorfer [100, Theorem 6].

Theorem 3.3.19 Let $l \in \{1, 2, \ldots, n-1\}$ be given. Furthermore, assume that
$$\hat{c} \in \bigl(\cos(l\pi/n), \cos((l-1)\pi/n)\bigr].$$
Then, there exist precisely $n - l$ distinct numbers
$$\hat{b}_j \in (-1, \cos(j\pi/n)), \quad j = 1, 2, \ldots, n - l,$$
such that the Chebyshev polynomial with respect to $\hat{E}_j := [-1, \hat{b}_j] \cup [\hat{c}, 1]$, $j = 1, 2, \ldots, n - l$, has $n + 2$ extremal points.

3.4 Computing an Asymptotic Convergence Factor

If the ratio ρ/K is **not** rational, the corresponding Chebyshev polynomial and the optimal polynomial are, in general, not explicitly known. Here, de Boor and Rice [30] formulated a *Remez type algorithm* for the numerical computation of the optimal polynomial. For an actual implementation see Sauer [110]. Also, Fischer and Modersitzki [44] developed a MATLAB package, which is capable of computing optimal polynomials, even for regions in the complex plane. It is publically available via the NETLIB facility.

3.4 Computing an Asymptotic Convergence Factor

In this section we devise a scheme for computing the asymptotic convergence factor (cf. (3.1.6))

$$\kappa(E) = \exp(-g(0; E^c, \infty))$$

associated with the set $E = [a, b] \cup [c, d]$, $a < b < 0 < c < d$. Note that such a set may contain the eigenvalues of a symmetric indefinite matrix. Here, g is the Green's function for E^c with pole at infinity. We remark that a slightly different scheme can be found in Freund [50].

It is not hard to verify, that the convergence factor is essentially invariant under linear transformations. To be precise, let l denote the linear transformation (cf.(3.2.1) and (3.2.2))

$$l(t) := \frac{a + d - 2t}{a - d}, \quad \xi := l(0) = \frac{a+d}{a-d},$$

which maps E onto $\hat{E} = [-1, \hat{b}] \cup [\hat{c}, 1]$. Then, it holds

(3.4.1) $$\kappa(E) = \kappa(\hat{E}; \xi) = \exp(-g(\xi; \hat{E}^c, \infty)).$$

In Theorem 3.3.18 we worked out an explicit representation for the Green's function in question. This, together with (3.4.1), yields

---

Corollary 3.4.1 *Let $\{\hat{E}, \hat{b}, \hat{c}, k, \rho\}$ (cf. (3.3.15)) be given. Then, the asymptotic convergence factor (3.4.1) is given by*

$$\kappa(\hat{E}; \xi) = \left| \frac{H(u_\xi - \rho; k)}{H(u_\xi + \rho; k)} \right|, \quad \xi = \varphi(u_\xi; \rho),$$

where the Theta function H is defined in (3.3.11) and the mapping φ is given in (3.3.17).

Hence, to compute the asymptotic convergence factor we have to get a handle on ρ and u_ξ. Notice that Lemma 3.3.13(c) implies that $u_\xi = \hat{u}_\xi + iK'$ is located in the segment $[-K + iK', iK']$. We have in view of (3.3.16) and Lemma 3.3.13(d)

$$\text{(3.4.2)} \qquad \rho = -\operatorname{sn}^{-1}(w_\rho), \quad \text{where} \quad w_\rho = \sqrt{\frac{1-\hat{b}}{2}},$$

and

$$\text{(3.4.3)} \qquad \hat{u}_\xi = -\operatorname{sn}^{-1}(w_\xi), \quad \text{where} \quad w_\xi = -\frac{1}{k \operatorname{sn}(\rho)} \sqrt{\frac{\xi - \hat{b}}{\xi + 1}},$$

respectively. Moreover, we deduce from Lemma 3.3.12 that

$$\left| \frac{H(u_\xi - \rho)}{H(u_\xi + \rho)} \right| = \left| \frac{\theta(\hat{u}_\xi - \rho)}{\theta(\hat{u}_\xi + \rho)} \right|.$$

Thus, we need a scheme for the computation of the inverse function sn^{-1} and a scheme for the evaluation of the Theta function $\theta(t)$, $t \in \mathbb{R}$.

The Inverse of the Elliptic Sine

Recall from (3.3.7) that the inverse of the elliptic sine is given by

$$z = \operatorname{sn}^{-1}(w) = \int_0^w \frac{dx}{\sqrt{(1-x^2)(1-k^2x^2)}}$$

So, we have to evaluate an elliptic integral of the first kind. There are several ways outlined in the literature to accomplish this task. We follow Carlson [16], [17, pp. 279].

The first step is to apply the variable transformation $x = w/\sqrt{t+1}$ to obtain the standard form

$$\text{(3.4.4)} \qquad \operatorname{sn}^{-1}(w) = \frac{w}{2} \int_0^\infty \frac{dt}{\sqrt{(t+1-w^2)(t+1-k^2w^2)(t+1)}}$$

$$=: w \, R_F(1-w^2, 1-k^2w^2, 1).$$

The next lemma shows that the evaluation of R_F is surprisingly easy.

3.4 Computing an Asymptotic Convergence Factor

Lemma 3.4.2 *Let $x_0^2, y_0^2, z_0^2 \in \{z \in \mathbb{C} : \arg(z) \neq \pi\} \cup \{0\}$ be numbers in the complex plane cut along the nonpositive real axis and assume that at most one of them is 0. Furthermore, define the following iterates*

$$x_{n+1} = \frac{1}{2}\sqrt{(x_n + y_n)(x_n + z_n)}$$
$$y_{n+1} = \frac{1}{2}\sqrt{(y_n + x_n)(y_n + z_n)}$$
$$z_{n+1} = \frac{1}{2}\sqrt{(z_n + x_n)(z_n + y_n)}.$$

Then, we have

$$\lim_{n \to \infty} x_n = \lim_{n \to \infty} y_n = \lim_{n \to \infty} z_n = L,$$

and (cf. (3.4.4))

$$R_F(x_0^2, y_0^2, z_0^2) = \frac{1}{L}.$$

Moreover, it holds

$$x_n^2 - y_n^2 = \frac{1}{4^n}(x_0^2 - y_0^2) \quad \text{and} \quad y_n^2 - z_n^2 = \frac{1}{4^n}(y_0^2 - z_0^2).$$

The next lemma states that we may apply the scheme above for the computation of ρ and $\hat{u}_\xi$, respectively.

Lemma 3.4.3 *Let $\{\hat{E}, \hat{b}, \hat{c}, k, \rho\}$ and let $\xi \in (\hat{b}, \hat{c})$ be given.*

(a) The number w_ρ, defined by (3.4.2), satisfies $1 - w_\rho^2 > 0$.

(b) The number w_ξ, defined by (3.4.3), satisfies $1 - w_\xi^2 > 0$.

Proof. To prove (a) observe $1 - w_\rho^2 = \frac{1}{2}(1+\hat{b}) \geq 0$. For the verification of statement (b) note that $1-w_\xi^2 > 0$ if, and only if $k^2\text{sn}^2(\rho;k)(\xi+1)+\hat{b}-\xi > 0$. We make use of (3.3.8), (3.3.13), and (3.3.14) to obtain

$$\begin{aligned} k^2\text{sn}^2(\rho)(\xi+1) + \hat{b} - \xi &= \xi(k^2\text{sn}^2(\rho) - 1) + \hat{b} + k^2\text{sn}^2(\rho) \\ &> -\hat{c}\text{dn}^2(\rho) + \hat{b} + k^2\text{sn}^2(\rho) \\ &= 2(1 - \text{cn}^2(\rho) - \text{sn}^2(\rho)) \\ &= 0, \end{aligned}$$

which concludes the proof. □

MATLAB Implementation of SNINV

```
function [wR_F]=sn_inv(w,k,tol)
%
% computes the inverse of the elliptic sine sn(w;k)
%
% desired accuracy of the result: tol
%

%%initialize
    x=sqrt(1-w^2);
    y=sqrt(1-k^2*w^2);
    z=1;
    dxy=abs(x^2-y^2); dyz=abs(y^2-z^2);

%%iterate
    while (abs(x^2-y^2)/dxy > tol) & (abs(y^2-z^2)/dyz > tol)

        x_old=x;
        y_old=y;
        z_old=z;

        x=sqrt((x_old+y_old)*(x_old+z_old))/2;
        y=sqrt((y_old+x_old)*(y_old+z_old))/2;
        z=2*x*y/(x_old+y_old);

    end %while

    wR_F=w/z;

    return;
```

3.4 Computing an Asymptotic Convergence Factor

Evaluation of a Theta Function

In this section we show how to compute $\theta(t)$, $t \in \mathbb{R}$, up to a prescribed relative error. We follow the derivation in an earlier version of Freund [50]. The idea is to approximate the infinite series (cf. (3.3.11))

$$\theta(t) = \vartheta_4(x;q) = 1 + 2\sum_{j=1}^{\infty}(-1)^j q^{j^2}\cos(2jx), \quad x = \frac{t}{2K}\pi,$$

by a finite one

(3.4.5) $$\theta^{(J)}(t) := 1 + 2\sum_{j=1}^{J}(-1)^j q^{j^2}\cos(2jx).$$

where $q = \exp(-\pi K'/K) < 1$.

We will make use of the following two facts, which can be found, for example, in Whittaker and Watson [133, pp. 462]

$$\vartheta_4(\pi/2;q) = \sqrt{2K/\pi}$$

and

$$\vartheta_4(x;q) \geq \vartheta_4(0;q) = \sqrt{2k'K/\pi}, \quad \text{for all} \quad x \in \mathbb{R}.$$

Then a straightforward analysis shows that

$$\begin{aligned}|\theta(t) - \theta^{(J)}(t)| &= 2\left|\sum_{j=J+1}^{\infty}(-1)^j q^{j^2}\cos(2jx)\right| \\ &\leq 2\sum_{j=J+1}^{\infty} q^{j^2} \\ &\leq 2q^{(J+1)^2}\sum_{j=0}^{\infty} q^{j^2} \\ &= q^{(J+1)^2}(1 + \vartheta_4(\pi/2;q)) \\ &= q^{(J+1)^2}(1 + \sqrt{2K/\pi}).\end{aligned}$$

This leads to

$$\left|\frac{\theta(t) - \theta^{(J)}(t)}{\theta(t)}\right| \leq q^{(J+1)^2}\frac{1 + \sqrt{\pi/(2K)}}{\sqrt{k'}}.$$

> **Lemma 3.4.4** Let $\varepsilon > 0$ and $k \in (0,1)$, with $k' = \sqrt{1-k^2}$, and $q = \exp(-\pi K'/K)$ be given. If J is the integer part of $|\log(\varepsilon\sqrt{k'}) + \log(1 + \sqrt{\pi/(2K)})|/|\log(q)|$, then
> $$\left|\frac{\theta(t) - \theta^{(J)}(t)}{\theta(t)}\right| \leq \varepsilon, \quad t \in \mathbb{R}.$$

MATLAB Implementation of ASYMPFAC

```
function [kappa]=asympfac(a,b,c,d,tol)
%
% computes the asymptotic convergence factor
% relative to E = [a,b] u [c,d]
%
% desired accuracy of the result: tol
%
% uses: sn_inv, JTheta_4
%
%%normalize E
   b_hat=(2*b-a-d)/(d-a);
   c_hat=(2*c-a-d)/(d-a);
   xi   =-(d+a)/(d-a);

%%compute k and rho
   k=sqrt(2*(c_hat-b_hat)/((1-b_hat)*(1+c_hat)));
   w_rho=sqrt((1-b_hat)/2);
   rho=-sn_inv(w_rho,k,tol);

%%complete elliptic integrals
   K=ellipk(1,k^2);
   K_prime=ellipk(1,1-k^2);

%%compute u_hat
   sn_rho=ellipj(rho,k^2);
   w_xi=-sqrt((xi-b_hat)/(xi+1))/(k*sn_rho);
   u_hat=-sn_inv(w_xi,k,tol);

%%compute kappa
   kappa=abs(JTheta_4((u_hat-rho),K,K_prime,tol)/...
             JTheta_4((u_hat+rho),K,K_prime,tol));

return;
```

3.4 Computing an Asymptotic Convergence Factor

Let us finish with a little illustrating example.

Example 3.4.5 Consider the family of sets $E(t) := [-1.8+t, -1.3+t] \cup (0+t, 0.2+t]$ for $t \in (0, .8]$. The next figure shows the asymptotic convergence rate of $E(t)$ as a function of t.

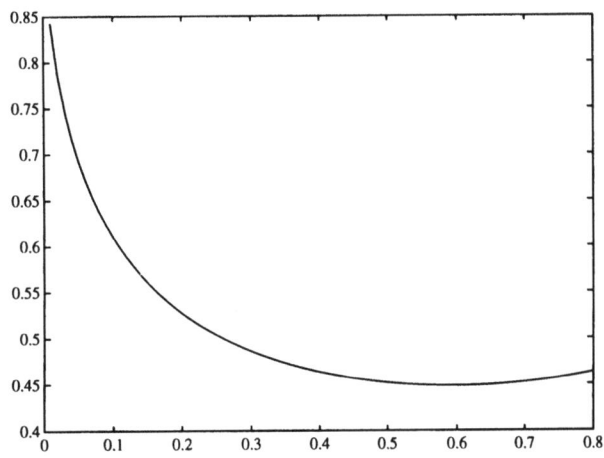

Figure 3.4.6 Asymptotic convergence rate of $E(t)$ versus t.

It turns out that, despite popular believe, the best asymptotic convergence rate is **not** achieved for the case where the origin is located in the middle of the two interior boundaries, i.e., for $t = 0.65$. The best rate is obtained for $t = 0.61\ldots$ □

4 Orthogonal Polynomials and Krylov Subspaces

In the introduction (cf. Section 1.1) we already indicated that there is a strong connection between the Krylov subspace $\mathcal{K}_n(A; r_0)$ and the space Π_{n-1} of all polynomials of degree not exceeding $n-1$

$$\mathcal{K}_n(A; r_0) = \{p(A)r_0 : p \in \Pi_{n-1}\}.$$

In particular, any basis of the polynomial space generates a basis for the Krylov subspace. In this section we will discuss two distinguished examples of this connection.

4.1 Generating a Basis; Orthonormal Case

Before we start, recall that (cf. 1.1.5)

$$L = \dim \mathcal{K}_N(A, r_0), \quad \text{with} \quad A \in \mathbb{R}^{N \times N},$$

is the maximum number of linearly independent vectors in $\mathcal{K}_N$.

Let $\psi_0, \psi_1, \ldots, \psi_{L-1}$ denote the orthonormal polynomials with respect to $\langle \cdot, \cdot \rangle$. These polynomials generate basis vectors

$$v_j := \psi_{j-1}(A)r_0, \quad 1 \leq j \leq L,$$

for the Krylov subspaces

(4.1.1) $$\mathcal{K}_n(A; r_0) = \text{span}\{v_1, v_2, \ldots, v_n\}, \quad n = 1, 2, \ldots, L.$$

Observe, that the basis vectors are orthonormal

(4.1.2) $$\langle v_j, v_k \rangle = \langle \psi_{j-1}, \psi_{k-1} \rangle = 0, \ j \neq k; \quad \|v_j\|^2 = \langle \psi_{j-1}, \psi_{j-1} \rangle = 1,$$

with respect to the (vector) inner product induced by the (polynomial) inner product $\langle \cdot, \cdot \rangle$ (see Section 1.1).

4.1 Generating a Basis; Orthonormal Case

The basis (4.1.1) has outstanding properties. To begin with, we note that the v_j's fulfill a three-term recurrence relation and may be computed by a Stieltjes-like algorithm (compare Algorithm 2.1.3).

Algorithm 4.1.1 *Let $n \leq L$ and let r_0 be given. The vectors $\{v_j\}_{j=1}^n$, orthonormal with respect to $\langle \cdot, \cdot \rangle$, may be computed by the following recursive procedure.*

Set
$$v_0 := 0, \quad \nu_0 = \langle 1, 1 \rangle, \quad \beta_1 := \sqrt{\nu_0}, \quad v_1 = r_0/\beta_1.$$

For $j = 1, 2, \ldots, n-1$ compute
$$\alpha_j = \langle Av_j, v_j \rangle, \quad \hat{v}_{j+1} = (A - \alpha_j I_N)v_j - \beta_j v_{j-1},$$
$$\beta_{j+1} = \|\hat{v}_{j+1}\|, \quad v_{j+1} = \hat{v}_{j+1}/\beta_{j+1}.$$

The grade L of r_0 is, in general, not known beforehand. Also, there may exist more than L (even infinitely many) orthonormal polynomials ψ_j with respect to $\langle \cdot, \cdot \rangle$. We stress that Algorithm 4.1.1 does not break down, as long as ψ_j's exist, since this implies

$$\beta_{j+1} = \|\hat{v}_{j+1}\| = \sqrt{\langle \hat{\psi}_j, \hat{\psi}_j \rangle} > 0.$$

Of course, the vectors v_j, $j > L$, are no longer linearly independent of the previous ones.

Let us collect together the vectors v_j in the $(N \times n)$ matrix

(4.1.3) $$V_n := [v_1 \quad v_2 \quad \cdots \quad v_n].$$

It should come as no surprise, that in accordance with (2.1.27) and (2.1.28) it holds

(4.1.4) $$AV_n = V_n J_n + \beta_{n+1} v_{n+1} e_n^T$$

or, equivalently,

(4.1.5) $$AV_n = V_{n+1} \hat{J}_n,$$

where J_n and $\hat{J}_n$, respectively, denote the Jacobi matrices with respect to $\langle \cdot, \cdot \rangle$. Also, we have by (4.1.2) and (4.1.4)

(4.1.6) $$\langle V_n, V_n \rangle = I_n \quad \text{and} \quad \langle V_n, AV_n \rangle = J_n,$$

i.e., Algorithm 4.1.1 may be viewed as a tridiagonalization procedure.

There is yet another important fact we would like to point out. We know that the orthonormal polynomials ψ_n are uniquely characterized as the solution of a certain L_2 approximation problem (cf. Theorem 2.1.4). This in particular implies that the ψ_n's are invariant under translation. Consequently, Algorithm 4.1.1 produces the same sequence of vectors v_n from r_0 if $A - \rho I_N$ is used in place of A for any number ρ. Hence, Algorithm 4.1.1 does not suffer from A being badly conditioned. In particular, it is as stable when applied to indefinite matrices as it is for positive definite ones.

Lanczos Method

Let us now apply Algorithm 4.1.1 to the inner product (cf. (1.1.12))

$$\langle p, q \rangle_{\text{GAL}} = r_0^{\text{T}} p(A) q(A) r_0.$$

Recall that this inner product induces the Euclidean inner product on the Krylov subspace $\mathcal{K}_n(A; r_0)$. For example, (4.1.6) reads

$$V_n^{\text{T}} V_n = I_n \quad \text{and} \quad V_n^{\text{T}} A V_n = J_n,$$

and, most important, Algorithm 4.1.1 converts into the celebrated *Lanczos method* (cf. Lanczos[87]) with

$$\alpha_j = v_j^{\text{T}} A v_j \quad \text{and} \quad \beta_{j+1} = \|\hat{v}_{j+1}\|_2.$$

This time there exist only L orthogonal polynomials with (compare the comment after (2.1.7))

$$\beta_{L+1} = \|\hat{v}_{L+1}\|_2 = \sqrt{\langle \hat{\psi}_L, \hat{\psi}_L \rangle_{\text{GAL}}} = 0.$$

MATLAB Implementation of LANCZOS

In finite precision arithmetic, the stopping criteria $\beta_{L+1} = 0$ is not practicable. Instead one may stop the Lanczos process if β_j is sufficiently small.

4.2 Generating a Basis; Monic Case

```
function [V,alpha,beta]=LANCZOS(A,r0,n,tol)
%
% computes an orthonormal basis of the Krylov subspace
%
%     K_n(A,r0) = span{v_1,v_2,...,v_n}
%
% and the entries of the Jacobi matrix J_n
%
% stops if beta_j < tol
%
% uses: matvec
%
%%initialize
    j=1; N=length(r0);
    v_old=zeros(N,1); v_hat=r0;
    beta(j)=norm(v_hat);

%%iterate
    while (j < n+1) & (beta(j) > tol)

      v=v_hat/beta(j); Av=matvec(A,v); alpha(j)=v'*Av;
      v_hat=Av-alpha(j)*v-beta(j)*v_old;
      j=j+1; beta(j)=norm(v_hat);
      v_old=v; V=[V,v];

    end; %while

return;
```

4.2 Generating a Basis; Monic Case

One may of course generate a basis for the Krylov subspace with the help of any set of linearly independent polynomials. In view of Algorithm 2.5.6 and Algorithm 2.6.6 we will have occasion to employ a basis

$$(4.2.1) \qquad \text{span}\{v_1^{\text{MO}}, v_2^{\text{MO}}, \ldots, v_n^{\text{MO}}\} = \mathcal{K}_n(A, r_0), \quad v_j^{\text{MO}} := \psi_{j-1}^{\text{MO}}(A) r_0,$$

generated by the monic polynomials ψ_n^{MO} that are orthogonal with respect to the modified inner product $\langle \cdot, \cdot t^2 \rangle$. In this section we briefly outline the corresponding procedure.

The three-term recurrence relation for the monic polynomials (cf. (2.1.19)) directly translates to an update formula for the basis elements

$$\begin{aligned} v_{n+1}^{\text{MO}} &= \psi_n^{\text{MO}}(A) r_0 \\ &= (A - \alpha_n) \psi_{n-1}^{\text{MO}}(A) r_0 - \beta_n \psi_{n-2}^{\text{MO}}(A) r_0 \\ &= A v_n^{\text{MO}} - \alpha_n v_n^{\text{MO}} - \beta_n v_{n-1}^{\text{MO}}, \end{aligned}$$

where the coefficients are given by (cf. (2.1.20))

$$\alpha_n = \frac{\langle t\psi_{n-1}^{\text{MO}}, t^2\psi_{n-1}^{\text{MO}}\rangle}{\langle \psi_{n-1}^{\text{MO}}, t^2\psi_{n-1}^{\text{MO}}\rangle} \quad \text{and} \quad \beta_n = \frac{\langle \psi_{n-1}^{\text{MO}}, t^2\psi_{n-1}^{\text{MO}}\rangle}{\langle \psi_{n-2}^{\text{MO}}, t^2\psi_{n-2}^{\text{MO}}\rangle}.$$

Again, the special inner product $\langle \cdot, \cdot t^2\rangle_{\text{GAL}}$ leads to a particularly convenient expression for these coefficients. Here, we have

$$\alpha_n = \frac{(Av_n^{\text{MO}})^{\text{T}} A(Av_n^{\text{MO}})}{(Av_n^{\text{MO}})^{\text{T}} Av_n^{\text{MO}}} \quad \text{and} \quad \beta_n = \frac{(Av_n^{\text{MO}})^{\text{T}} Av_n^{\text{MO}}}{(Av_{n-1}^{\text{MO}})^{\text{T}} Av_{n-1}^{\text{MO}}}.$$

Algorithm 4.2.1 *Let $n \leq L$ and let r_0 be given. The vectors $\{v_j^{\text{MO}}\}_{j=1}^n$, orthogonal with respect to $\langle \cdot, \cdot t^2\rangle_{\text{GAL}}$, may be computed by the following recursive procedure.*

Set
$$v_0^{\text{MO}} := 0, \quad \beta_1 := 0, \quad v_1^{\text{MO}} = r_0.$$

For $j = 1, 2, \ldots, n-1$ compute

$$\alpha_j = \frac{(Av_j^{\text{MO}})^{\text{T}} A(Av_j^{\text{MO}})}{(Av_j^{\text{MO}})^{\text{T}}(Av_j^{\text{MO}})}, \quad v_{j+1}^{\text{MO}} = (A - \alpha_j I_N)v_j^{\text{MO}} - \beta_j v_{j-1}^{\text{MO}},$$

$$\beta_{j+1} = \frac{(Av_{j+1}^{\text{MO}})^{\text{T}} A(Av_{j+1}^{\text{MO}})}{(Av_j^{\text{MO}})^{\text{T}}(Av_j^{\text{MO}})}.$$

5 Estimating the Spectrum and the Distribution function

The parameter dependent schemes require some a priori information about the underlying scheme. In this chapter we show how to estimate the spectrum of a given symmetric indefinite matrix and how to approximate its eigenvalue distribution.

To test the devised schemes we first define a class of "model matrices".

5.1 The Model Problem

Let $\Omega := \{(x, y) : 0 < x, y < 1\}$ denote the unit square and Γ its boundary. We consider the Helmholtz equation

$$
\begin{aligned}
-\Delta u(x,y) - \tau u(x,y) &= f(x,y), \quad (x,y) \in \Omega, \\
u(x,y) &= g(x,y), \quad (x,y) \in \Gamma,
\end{aligned}
\tag{5.1.1}
$$

on the unit square with Dirichlet boundary conditions. Here, f and g are given functions and $\tau \geq 0$ is a non-negative real parameter. We discretize this selfadjoint elliptic boundary value problem using 5 point centered differences on a uniform grid with mesh size $h = 1/(m+1)$. This results in the linear system [†]

$$
A(\tau, m) x = f \tag{5.1.2}
$$

with the $(m^2 \times m^2)$ block tridiagonal coefficient matrix $A(\tau, m)$ given by

$$
A(\tau, m) = \begin{bmatrix}
C_m & -I_m & 0 & \cdots & 0 \\
-I_m & C_m & \ddots & \ddots & \vdots \\
0 & \ddots & \ddots & \ddots & 0 \\
\vdots & \ddots & \ddots & \ddots & -I_m \\
0 & \cdots & 0 & -I_m & C_m
\end{bmatrix}
$$

[†] For convenience we use the letter f for both the function $f(x, y)$ and the right hand side vector f.

involving the $(m \times m)$ tridiagonal blocks

$$C_m = \begin{bmatrix} 4 - \tau h^2 & -1 & 0 & \cdots & 0 \\ -1 & 4 - \tau h^2 & \ddots & \ddots & \vdots \\ 0 & \ddots & \ddots & \ddots & 0 \\ \vdots & \ddots & \ddots & \ddots & -1 \\ 0 & \cdots & 0 & -1 & 4 - \tau h^2 \end{bmatrix}.$$

This set of equations is often used for testing purposes and is sometimes referred to as the "model problem". We stress, that we view (5.1.1) as a source for producing test matrices. Our goal is to solve the linear system (5.1.2) rather than to approximate the solution of (5.1.1). In particular, we may choose parameters τ which have no "physical meaning", but produce matrices $A(\tau, m)$ with some wanted properties.

However, the spectrum of $A(\tau, m)$ is explicitly known (Varga [126])

$$\sigma(A(\tau, m)) = \left\{ 4 - \tau h^2 + 2 \cos \frac{k\pi}{m+1} + 2 \cos \frac{l\pi}{m+1} \; : \; 1 \le k, l \le m \right\}.$$

So, $A(\tau, m)$ is indefinite for

$$\frac{4}{h^2}(1 - \cos \pi h) < \tau < \frac{4}{h^2}(1 + \cos \pi h), \quad h = \frac{1}{m+1}.$$

Example 5.1.1

Let us have a closer look at the matrix $A(350, 25)$. Its eigenvalues (vertical lines) are plotted in Figure 5.1.2, which constitutes a good test problem for Postscript printers.

In Figure 5.1.3 we display a particular eigenvalue distribution function for $A(350, 25)$ (cf. Example 2.1.2)

$$\sigma(t) = \sum_{j=1}^{L} \sigma_j^2 \delta(t - \lambda_j),$$

where $\sigma_j^2 := \#\lambda_j/625$ is defined as scaled multiplicity $\#\lambda_j$ of the eigenvalue λ_j.

5.1 The Model Problem

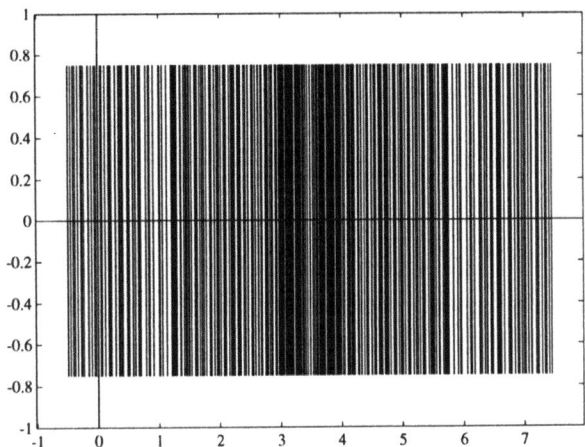

Figure 5.1.2 Continuous vertical lines: eigenvalues of $A(350, 25)$.

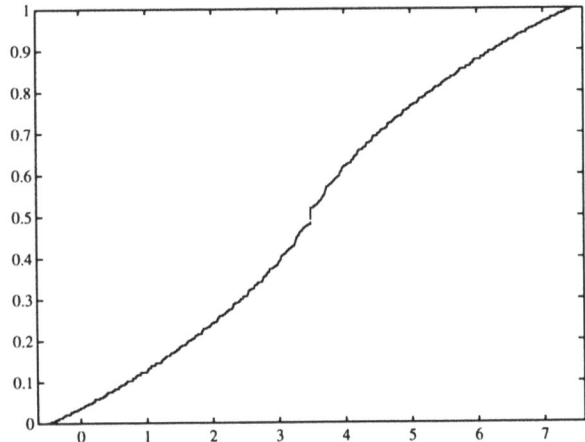

Figure 5.1.3 Distribution function $\sigma(t)$ for $A(350, 25)$.

$A(350, 25)$ has 24 eigenvalues in $[-.489, -.00019]$ and 601 eigenvalues in $[0.0433, 7.46]$, respectively. Moreover, $A(350, 25)$ has $L = 400$ distinct eigenvalues. □

5.2 Estimating the Spectrum

Let A be given symmetric indefinite matrix; i.e., the spectrum $\sigma(A)$ of A is located in the union of two disjoint intervals

$$\sigma(A) \subset E := [a,b] \cup [c,d], \ a < b < 0 < c < d.$$

In this section we outline an approach for the estimation of the boundary points a, b, c and d. The scheme is based on relations between the zeros of orthogonal residual polynomials, the zeros of kernel polynomials and the eigenvalues of the given matrix.

In this section we exclusively consider polynomials orthogonal with respect to the inner product $\langle \cdot, \cdot \rangle_{\text{GAL}}$ which is defined by the eigenvalues λ_j of A and by the expansion coefficients σ_j of the starting residual r_0 (cf. (1.1.6) and (1.1.11))

$$\langle p,q \rangle_{\text{GAL}} = r_0^T p(A) q(A) r_0 = \sum_{j=1}^{L} \sigma_j^2 p(\lambda_j) q(\lambda_j).$$

For a plot of the underlying distribution $\sigma(t)$ function, see Figure 2.1.1 or Figure 5.1.3, respectively. Again let us denote by ψ_n the orthonormal polynomials with respect to $\langle \cdot, \cdot \rangle_{\text{GAL}}$ (cf. (2.1.4))

$$\langle \psi_i, \psi_j \rangle_{\text{GAL}} = 0, \quad \text{for} \ i \neq j,$$

and by K_n the kernel polynomials with respect to $\langle \cdot, \cdot \rangle_{\text{GAL}}$ and the parameter $\xi = 0$ (cf. (2.5.1))

$$\langle K_i, t K_j \rangle_{\text{GAL}} = 0, \quad \text{for} \ i \neq j.$$

The point is that in view of (2.1.7) the eigenvalues of A are just the zeros of the orthogonal polynomial of degree L

$$\psi_L(t) = \prod_{j=1}^{L} (t - \lambda_j).$$

In other words, ψ_L is the minimal polynomial of A with respect to r_0

$$\psi_L(A) r_0 = 0.$$

So, there is a close connection between the zeros of the ψ_n's and the eigenvalues of A. From Theorem 2.1.6 we know that the zeros $\theta_j^{(n)}$ of ψ_n are

5.2 Estimating the Spectrum

just the eigenvalues of the associated Jacobi matrix J_n with corresponding eigenvectors $y_j^{(n)} := \Psi_{n-1}(\theta_j^{(n)})$. Furthermore, in Section 4.1 we showed that

$$V_n^T A V_n = J_n, \quad V_n^T V_n = I_n,$$

where

$$V_n = [v_1 \quad v_2 \quad \cdots \quad v_n] \quad \text{and} \quad v_j = \psi_{j-1}(A) r_0.$$

Hence, the pairs $(\theta_j^{(n)}, V_n y_j^{(n)})$ can be seen as *Rayleigh-Ritz approximations* to eigenpairs of A from the Krylov subspace $\mathcal{K}_n(A; r_0)$. In this "language" the eigenvalues of J_n, which are nothing but the eigenvalues of the orthogonal projection of A onto $\mathcal{K}_n(A; r_0)$, are called *Ritz values* (see, Parlett [97, §11.3]).

Next, we state an important corollary to Theorem 2.1.5 (compare van der Sluis and van der Vorst [124, §2], [125, §2]).

Corollary 5.2.1 *Let A be a symmetric indefinite matrix with all eigenvalues located in $E = [a, b] \cup [c, d]$ and let $\theta_1^{(n)} < \theta_2^{(n)} < \cdots < \theta_n^{(n)}$ denote the Ritz values of A.*

(a) *Let k be fixed. Then $\theta_k^{(n)}$ decreases and $\theta_{n-k}^{(n)}$ increases as a function of the dimension n.*

(b) *Between two Ritz values $\theta_j^{(n)}$ and $\theta_{j+1}^{(n)}$ there is at least one eigenvalue.*

(c) *There is at most one Ritz value in the gap $[b, c]$ and no Ritz value outside $[a, d]$.*

The next corollary says that the Lanczos process is shift invariant. It is a direct consequence of Theorem 2.1.4.

Corollary 5.2.2 *Let A be a symmetric matrix with Rayleigh-Ritz approximations $(\theta_j^{(n)}, V_n y_j^{(n)})$. Furthermore let $\rho \in \mathbb{R}$ be a given number.*

(a) *If A is changed to $A + \rho I_N$, and r_0 is left unchanged, the Ritz values change to $\theta_j^{(n)} + \rho$.*

(b) *If A is changed to ρA, and r_0 is left unchanged, the Ritz values change to $\rho \theta_j^{(n)}$.*

(c) *The Ritz vectors $V_n y_j^{(n)}$ do not change under the transformations in (a) and (b), respectively.*

If one would like to use the Ritz values to approximate the spectrum of a given symmetric indefinite matrix, then the fact that there could be one Ritz value in the gap $[b, c]$ may cause some trouble.

Next, we discuss the roots of the kernel polynomials. They may also be seen as a certain Rayleigh-Ritz approximation. More precisely, they are the reciprocals of the Ritz values of the orthogonal projection of A^{-1} onto $A\mathcal{K}_n(A; r_0)$ (see Manteuffel and Otto [89]; and Paige, Parlett, and van der Vorst[93]). These special Ritz values have several names in the literature. Freund [48] calls them pseudo-Ritz values, Manteuffel and Otto call them generalized Ritz values, whereas Paige, Parlett, and van der Vorst call them *harmonic Ritz values*. We will use the last designation.

The next corollary states one of their main properties. It follows directly from Theorem 2.5.8, were we investigated zeros of kernel polynomials with respect to an abitrary inner product.

Corollary 5.2.3 *Let A be a symmetric indefinite matrix with all eigenvalues located in $E = [a, b] \cup [c, d]$. Then, there is at most one harmonic Ritz value outside $[a, d]$ and no harmonic Ritz value in the gap $[b, c]$.*

5.2 Estimating the Spectrum

We remark, that the zeros of kernel polynomials are definitely not invariant under translation, since their definition (cf. (2.5.1)) involves the translation dependent parameter ξ.

However, to estimate the extreme values a, b, c, and d of the spectrum $E = [a, b] \cup [c, d]$ it is advisable to compute both the Ritz values $\theta_j^{(n)}$ and the harmonic Ritz values $\hat{\theta}_j^{(n)}$. After having done this, one may perform a "discrete intersection" of these two sets in order to get rid of possible Ritz values in the gap $[b, c]$ and of possible harmonic Ritz values outside the spectrum $[a, d]$. To be precise, let

$$\theta_1^{(n)} < \theta_2^{(n)} < \cdots < \theta_n^{(n)}$$

and

$$\hat{\theta}_1^{(n)} < \hat{\theta}_2^{(n)} < \cdots < \hat{\theta}_k^{(n)} < 0 < \hat{\theta}_{k+1}^{(n)} < \hat{\theta}_{k+2}^{(n)} < \cdots < \hat{\theta}_n^{(n)}.$$

denote Ritz values and harmonic Ritz values, respectively, for a given matrix. Then, Cor. 5.2.1 and Cor. 5.2.3 imply that

(5.2.1) $$a \leq \theta_1^{(n)} < \hat{\theta}_k^{(n)} \leq b < c \leq \hat{\theta}_{k+1}^{(n)} < \theta_n^{(n)} < d,$$

under the assumption that $\theta_1^{(n)} < 0$ and that $k > 0$.

It remains to discuss the rate of convergence of the Ritz values and harmonic Ritz values to "their eigenvalues". Results in this direction are known under the name *Kaniel-Paige theory*. For details we refer to Golub and van Loan [71, §9.1.4], to Parlett [97, §13], and to Paige, Parlett, and van der Vorst [93]. We will not discuss them here. Instead we will just present a numerical example that gives an idea of "what is going on".

Example 5.2.4 Here A is a modified Helmholtz matrix $A(350, 25)$ (cf. (5.1.2)) of order $N = 625$. Since the matrix $A(350, 25)$ has all its eigenvalues located in $[-.489, -.00019] \cup [0.0433, 7.46]$ (cf. Example 5.1.1), the gap $[-.00019, 0.0433]$ is not well suited for "plotting purposes". Therefore we "artificially enlarged the gap" by shifting the negative eigenvalues by $-.5$. This leads to a matrix A with its spectrum contained in $[-.989, -.50019] \cup [0.0433, 7.46] =: [a, b] \cup [c, d]$. A random vector with components uniformly distributed in $[-1, 1]$ was selected as initial approximation. The right hand side f was chosen such that $x_* = (1, 1, \ldots, 1)$ is the solution.

We applied the Lanczos method, starting with $r_0 = f - Ax_0$, to A. This produced the entries of the Jacobi matrix J_n. Finally, the Ritz values and harmonic Ritz values were computed by solving the eigenvalue problems as outlined in Theorem 2.1.6 and Theorem 2.5.11, respectively.

The next figure illustrates some results.

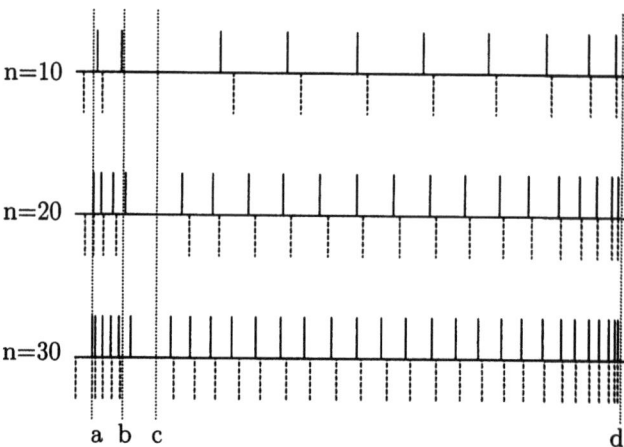

Figure 5.2.5 Vertical solid lines: Ritz values; vertical dashed lines: harmonic Ritz values; vertical dotted lines: extreme values of the spectrum.

We computed the two sets $\{\theta_j^{(n)}\}_{j=1}^n$ and $\{\hat{\theta}_j^{(n)}\}_{j=1}^n$ for the sample values $n = 10, 20, 30$. It is apparent that in each case the smallest harmonic Ritz value is to the left of the spectrum $\hat{\theta}_1^{(n)} < a$. Furthermore, in the cases $n = 20, 30$ one Ritz value is located in the gap $[b, c]$. Also, the boundary points a, b, d are pretty well approximated by both sets of estimates. Only the point c causes some trouble. It is a typical phenomenon for the Rayleigh-Ritz approximations, that interior eigenvalues are much harder to estimate than the extreme eigenvalues. □

5.3 Approximating the Distribution Function

In this section we show how to approximate the eigenvalue distribution of a symmetric indefinite matrix. The approach is based on the Lanczos process. We point out that spectral estimation based on the Lanczos process is a widely used technique in applications (see, e.g., Reinhardt [105] and the

5.3 Approximating the Distribution Function

references given there). Also, Fischer and Freund [39] applied this technique to the design of polynomial preconditioners for Hermitian positive systems.

Lanczos Method and Distribution Functions

The problem is to approximate the distribution function (cf. Example 2.1.2)

$$\sigma(t) = \sum_{j=1}^{L} \sigma_j^2 \delta(t - \lambda_j),$$

defined by the eigenvalues λ_j of A and by the coefficients σ_j of the eigenvector expansion of the starting residual r_0 (cf. (1.1.6)). Recall, that the inner product

$$\langle p, q \rangle_{\text{GAL}} = r_0^T p(A) q(A) r_0 = \sum_{j=1}^{L} \sigma_j^2 p(\lambda_j) q(\lambda_j) = \int_a^d p(t) q(t) d\sigma(t)$$

is given by this distribution function.

To estimate $\sigma(t)$, we make use of the connection between moments and distribution functions as stated in Theorem 2.2.5. More precisely, suppose we run the Lanczos process (cf. Section 4.1) for n steps. The process then has generated the entries of the Jacobi matrix J_n associated with $\langle \cdot, \cdot \rangle_{\text{GAL}}$. The associated Ritz values $\theta_j^{(n)}$ and the first component τ_j of the orthonormal eigenvectors of J_n define a new inner product (cf.(2.2.4))

$$(5.3.1) \qquad \langle p, q \rangle_n := \sum_{j=1}^{n} \left(\tau_j^{(n)}\right)^2 p(\theta_j^{(n)}) q(\theta_j^{(n)}) = \int_a^d p(t) q(t) d\tau(t)$$

with " low order" distribution function (cf.(2.2.5))

$$\tau(t) = \sum_{j=1}^{n} \left(\tau_j^{(n)}\right)^2 \delta(t - \theta_j^{(n)}).$$

The constructed inner product (5.3.1) can be seen as Gaussian quadrature formula with respect to $\langle \cdot, \cdot \rangle_{\text{GAL}}$ (cf. (2.2.4))

$$\langle p, 1 \rangle_n = \langle p, 1 \rangle_{\text{GAL}}, \quad \text{for all} \quad p \in \Pi_{2n-1}.$$

Consequently, in view of Theorem 2.2.5, the difference function $\tau(t) - \sigma(t)$ has precisely $2n - 1$ sign changes in the interval $[a, d]$.

Let us illustrate this property in the following example.

Example 5.3.1 Here, the Helmholtz matrix $A(350, 25)$ (cf. (5.1.2)) of order $N = 625$ served as the test-matrix. First, we computed $\sigma(t)$ with respect to $r_0 = f - Ax_0$, where $f = A(1, 1, \ldots, 1)^T$ and x_0 is a random vector with components uniformly distributed in $[-1, 1]$. We performed 10 Lanczos steps to obtain the following distribution function.

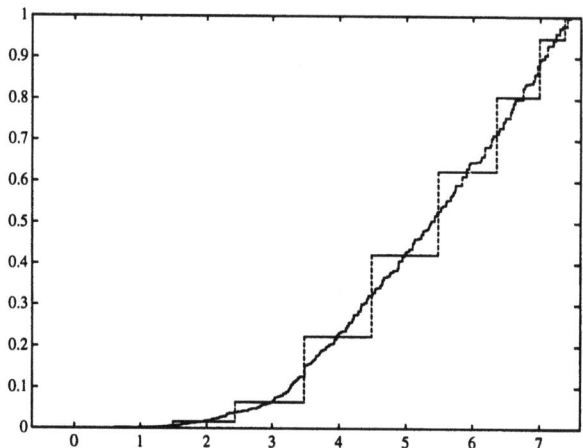

Figure 5.3.2 Distribution functions $\sigma(t)$ and $\tau(t)$ (for $n = 10$) with respect to $A(350, 25)$ and r_0.

Since the respective distribution functions are not very visible at all portions of the spectrum we did the same experiment but for a different initial vector $\hat{r}_0$. Let $\#\lambda_j$ denote the multiplicity of the eigenvalue λ_j. Then we constructed $\hat{r}_0$ such that the coefficients σ_j^2 in the distribution function $\sigma(t)$ display, up to a scaling factor, the multiplicity of λ_j (compare Figure 5.1.3)

$$\sigma_j^2 = \frac{\#\lambda_j}{N}.$$

Again, we performed 10 Lanczos steps to obtain the distribution function in the next figure.

Notice, that the midpoints of the vertical steps of $\tau(t)$ are very good approximations to points on the (unknown) curve $\sigma(t)$.

5.3 Approximating the Distribution Function

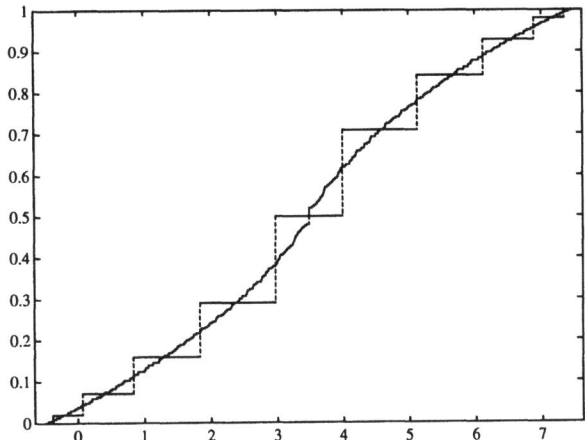

Figure 5.3.3 Distribution functions $\sigma(t)$ and $\tau(t)$ (for $n = 10$) with respect to $A(350, 25)$ and $\hat{r}_0$.

Recall, that the Chebyshev distribution function fulfills this "half step property" in a precise sense (cf. Example 2.2.3). □

Next, we outline an approach for estimating the unknown distribution function by an appropriate continuous function.

As is suggested by the figure above, we use interpolation at the vertical midpoints of $\tau(t)$ to construct an estimate, denoted by $s(t)$ in the sequel, for the true eigenvalue distribution $\sigma(t)$.

Before we state the interpolation problem we have to set up some boundary conditions. The Ritz values are strictly located in the interval given by the smallest and largest eigenvalue (cf. Theorem 2.1.5)

$$\lambda_1 < \theta_1^{(n)} < \theta_2^{(n)} < \cdots < \theta_n^{(n)} < \lambda_L, \quad n < L.$$

We introduce two "artificial Ritz values" by

$$\theta_0^{(n)} := \begin{cases} \dfrac{\theta_1^{(n)}}{2} & \text{for } \theta_1^{(n)} > 0 \\ \theta_1^{(n)} - \dfrac{\theta_2^{(n)} - \theta_1^{(n)}}{2} & \text{for } \theta_1^{(n)} \leq 0, \end{cases}$$

and
$$\theta_{n+1}^{(n)} := \begin{cases} \dfrac{\theta_n^{(n)}}{2} & \text{for } \theta_n^{(n)} < 0 \\ \theta_n^{(n)} + \dfrac{\theta_n^{(n)} - \theta_{n-1}^{(n)}}{2} & \text{for } \theta_n^{(n)} \geq 0. \end{cases}$$

Since, in general, the extreme Ritz values are good approximations to the extreme eigenvalues we use $\theta_0^{(n)}$ and $\theta_{n+1}^{(n)}$, respectively, to model the unknown distribution function $\sigma(t)$ outside the interval $[\lambda_1, \lambda_L]$. To this end, observe that
$$\sigma(t) = \tau(t) = 0 \quad \text{for} \quad t < \lambda_1,$$
and
$$\sigma(t) = \tau(t) = \nu_0 = \|r_0\|_2^2 \quad \text{for} \quad t \geq \lambda_L.$$

To justify the latter statement, we make use of $\psi_0(t) = \nu_0^{-1/2}$ and obtain
$$1 = \langle \psi_0, \psi_0 \rangle_{\text{GAL}} = \sum_{j=1}^L \sigma_j^2 \psi_0^2(\lambda_j) = \frac{1}{\nu_0} \sum_{j=1}^L \sigma_j^2 = \frac{1}{\nu_0} \sigma(\lambda_L).$$

The same argument applies to $\tau(t)$.

Altogether, we define the interpolation problem by setting
$$t_j := \theta_j^{(n)}, \quad j = 0, 1, \ldots, n+1$$
and
$$\vartheta_j := \begin{cases} 0 & \text{for } j = 0, \\ \tau_j^2/2 + \sum_{i=1}^{j-1} \tau_i^2 & \text{for } j = 1, 2, \cdots, n, \\ \nu_0 & \text{for } j = n+1. \end{cases}$$

Finally, the estimated distribution is then chosen as a monotone function $s \in C^1[-1, 1]$ satisfying

(5.3.2) $$s(t_j) = \vartheta_j, \quad j = 0, 1, \cdots, n+1.$$

We stress that in the indefinite case the above approach does not require any estimates for the interior eigenvalue boundaries, i.e., for the eigenvalues that define the gap including the origin. This is one of the strengths of this method.

5.3 Approximating the Distribution Function

Monotone Spline

One obvious choice for the interpolating function s is a monotone piecewise cubic interpolant. In this section, we briefly describe how to construct such a function. We follow the derivation of Fritsch and Carlson [55] and Fritsch and Butland [54].

We are looking for a function s with positive derivative on $[t_0, t_{n+1}]$. To achieve this goal, one "tunes" the derivative $d_j = s'(t_j)$ at the given points t_j, $j = 0, 1, \cdots, n+1$. To do this it turns out to be advantageous to express s in terms of the d_j's (compare [55]). We obtain for $t \in [t_j, t_{j+1}]$

$$s(t) = \left(\frac{d_j + d_{j+1} - 2\Delta_j}{h_j^2}\right)(t - t_j)^3$$
$$+ \left(\frac{-2d_j - d_{j+1} + 3\Delta_j}{h_j}\right)(t - t_j)^2 + d_j(t - t_j) + \vartheta_j,$$

where $h_j := t_{j+1} - t_j$ and $\Delta_j := (\vartheta_{j+1} - \vartheta_j)/h_j$, $j = 0, 1, \cdots, n$.

By construction, any choice of the free parameters d_j leads to a function $s \in C^1[-1, 1]$ that solves the interpolation problem (5.3.2). The remaining step is to adjust the d_j's to make s monotone on $[-1, 1]$.

Note that the d_j's are not uniquely determined, and various choices have been discussed in the literature. Here, we use a formula proposed by Brodlie [13] and Fritsch and Butland [54]
(5.3.3)
$$d_j := \begin{cases} \dfrac{\Delta_{j-1}\Delta_j}{\xi_j \Delta_j + (1 - \xi_j)\Delta_{j-1}} & \text{for } \Delta_{j-1}\Delta_j > 0, \\ 0 & \text{otherwise}, \end{cases} \quad j = 1, 2, \cdots, n,$$

where $\xi_j := (h_{j-1} + 2h_j)/(3(h_{j-1} + h_j))$. In addition to (5.3.3), we still need to choose the boundary conditions d_0 and d_{n+1}, respectively. Since, the Lanczos process does not offer any information about the derivatives at the endpoints, we select a weak version of the so-called "not-a-knot" condition (see de Boor [29, pp. 54]). Here one chooses d_0 and d_n such that s is twice continuously differentiable on $[t_0, t_2)$ and $(t_{n-1}, t_{n+1}]$, respectively. This leads to

(5.3.4)
$$d_0 = \frac{h_0}{h_1}(3\Delta_1 - (2d_1 + d_2)) + 3\Delta_0 - 2d_1,$$
$$d_{n+1} = \frac{h_n}{h_{n-1}}(3\Delta_{n-1} - (2d_n + d_{n-1})) + 3\Delta_n - 2d_n.$$

```
function [s]=MPCI(t,x,y)
%
% computes and evaluates a
% monotone piecewise cubic interpolant s(t)
%
% interpolation data:     (x,y)
% sample points:          t
%

%%initialize
   n=length(x);   d=zeros(n,1);  I=1:n-1;
   m=length(t);   s=zeros(m,1);
   h    =x(I+1)-x(I);
   delta=(y(I+1)-y(I))./h;  I(n-1)=[];
   xi   =(h(I)+2*h(I+1))./(3*(h(I)+h(I+1)));

%%derivatives at x_j
   for j=2:n-1
       if delta(j-1)*delta(j) > 0
          d(j) =(delta(j-1) * delta(j))/ ...
                (xi(j-1)*delta(j)+(1-xi(j-1))*delta(j-1));
       end;
   end;

%%"not-a-knot" condition
   d(1)=h(1)*(3*delta(2)-(2*d(2)+d(3)))/h(2)+...
        3*delta(1)-2*d(2);
   d(n)=h(n-1)*(3*delta(n-1)-(2*d(n-1)+d(n-2)))/h(n-2)+...
        3*delta(n-1)-2*d(n-1);
   if d(1) < 0,             d(1)=0; end;
   if d(1) > 3*delta(1),  d(1)=3*delta(1); end;
   if d(n) < 0,             d(n)=0; end;
   if d(n) > 3*delta(n-1), d(n)=3*delta(n-1); end;

%%evaluate spline
   I=find(t < x(1));
   if ~isempty(I)
       s(I)=y(1)*ones(length(I),1);
   end;
   for j=1:n-1
       I=find(t>=x(j) & t<x(j+1));
       if ~isempty(I)
          s(I)=((d(j)+d(j+1)-2*delta(j))/h(j)^2)*...
               (t(I)-x(j)).^3+((-2*d(j)-d(j+1)+...
               3*delta(j))/h(j))*(t(I)-x(j)).^2+...
               d(j)*(t(I)-x(j))+y(j);
       end;
   end;
   I=find(t >= x(n));
   if ~isempty(I)
       s(I)=y(n)*ones(length(I),1);
   end;
return;
```

5.3 Approximating the Distribution Function

This special choice of d_0 and d_{n+1}, however, does not necessarily produce a monotone function s on the subintervals $[t_0, t_1]$ and $[t_n, t_{n+1}]$, respectively. Note that the additional requirements $d_0 \in [0, 3\Delta_0]$ and $d_{n+1} \in [0, 3\Delta_n]$ will lead to a monotone function. Thus, if, e.g., d_0 (computed by (5.3.4)) turns out to be negative or bigger than $3\Delta_0$, we simply set $d_0 = 0$ or $d_0 = 3\Delta_0$, respectively.

We would like to mention that FORTRAN codes for the described procedures are available in NETLIB (PCHIP package).

MATLAB Implementation of MPCI

Here, we present a MATLAB implementation of a scheme for the computation and evaluation of a Monotone Piecewise Cubic Interpolant (MPCI).

Depending on the application, it might be more economical to write two single programs; one for the computation of the coefficients d_j and one for the actual evaluation of the generated function. Also, it is straightforward to devise a scheme for the computation of the derivative s'.

Example 5.3.4 In this example we applied MPCI to the second test-matrix of Example 5.3.1. For illustration purposes we display also the interpolation points.

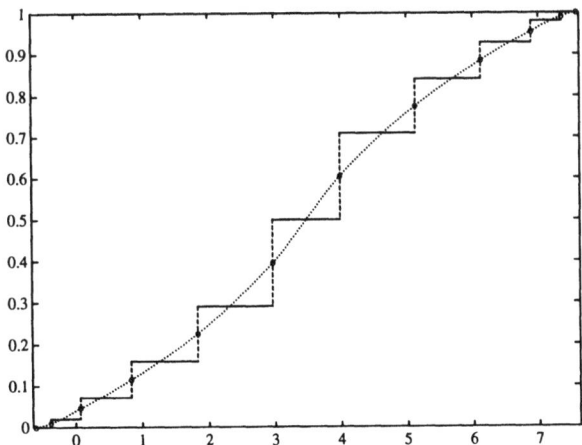

Figure 5.3.5 Continuous/dashed line: Distribution function $\tau(t)$ (for $n = 10$) with respect to $A(350, 25)$ and $\hat{r}_0$; dotted curve: monotone piecewise cubic interpolant.

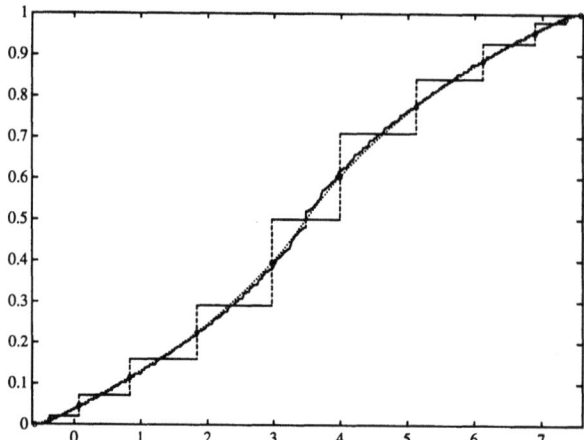

Figure 5.3.6 Continuous/dashed line: Distribution functions $\sigma(t)$ and $\tau(t)$ (for $n = 10$) with respect to $A(350, 25)$ and $\hat{r}_0$; dotted curve: monotone piecewise cubic interpolant.

In the second plot we displayed also the wanted distribution function $\sigma(t)$. Notice that the interpolant is almost indistinguishable from this curve. □

Computing New Orthogonal Polynomials

Let s be a monotone piecewise cubic interpolant with respect to the data (t_j, ϑ_j), where $t_0 < t_1 < \cdots < t_{n+1}$. Furthermore we denote by s_j the restriction of s onto the interval $[t_j, t_{j+1}]$. Note that $s_j \in \Pi_3$ is a cubic polynomial. The function s gives rise to the inner product

(5.3.5)
$$\begin{aligned}\langle p, q \rangle_s :&= \int_{t_0}^{t_{n+1}} p(t)q(t)ds(t) \\ &= \int_{t_0}^{t_{n+1}} p(t)q(t)s'(t)dt \\ &= \sum_{j=0}^{n} \int_{t_j}^{t_{j+1}} p(t)q(t)s'_j(t)dt.\end{aligned}$$

Clearly, there exists a set of orthonormal polynomials ψ_k^s with respect to $\langle \cdot, \cdot \rangle_s$. The computation of the three-term recurrence coefficients of these

5.3 Approximating the Distribution Function

polynomials may be accomplished by the Stieltjes procedure (cf. Algorithm 2.1.3). To do so, one has to be able to evaluate inner products of the form

(5.3.6) $$\langle t\psi_j^S, \psi_j^S \rangle_S \quad \text{and} \quad \langle \psi_j^S, \psi_j^S \rangle_S.$$

Note that in view of (5.3.5) this task comes down to the integration of polynomials. In principle, one only has to express the integrand in some polynomial basis and subsequently to integrate the basis elements. This approach may become unstable for high degree integrands, say $j = 100$ in (5.3.6).

In the following we will outline a scheme which turns out to be stable. This approach, however, has the disadvantage that one has to decide beforehand how many polynomials one likes to compute. To keep the issue of interest clear, let us just compute the orthonormal polynomials with respect to one single interval, i.e., we assume that

$$\langle p, q \rangle_S = \int_{t_0}^{t_1} p(t)q(t)s_0'(t)dt.$$

We are looking for the three-term recurrence coefficients in

$$\beta_{j+1}\psi_j^S(t) = (t - \alpha_j)\psi_{j-1}^S(t) - \beta_j \psi_{j-2}^S(t).$$

The approach is based on Gaussian quadrature. To this end, let J_m^L denote the Jacobi matrix associated with the *Legendre polynomials* shifted onto the interval $[t_0, t_1]$ (Szegö [122]) and let $\nu_0^L = (t_1 - t_0)$ denote the associated zero-order moment. We have by Theorem 2.2.1 and for $j \leq m-1$

(5.3.7)
$$\langle t\psi_j^S, \psi_j^S \rangle_S = \nu_0^L e_1^T \psi_j^S(J_m^L) J_m^L s_0'(J_m^L) \psi_j^S(J_m^L) e_1$$
$$\langle \psi_j^S, \psi_j^S \rangle_S = \nu_0^L e_1^T \psi_j^S(J_m^L) s_0'(J_m^L) \psi_j^S(J_m^L) e_1$$

Now, observe that with

$$z_j := \psi_j^S(J_m^L) e_1$$
$$= \frac{1}{\beta_{j+1}}\left((J_m^L - \alpha_j I_m)\psi_{j-1}^S(J_m^L)e_1 - \beta_j \psi_{j-2}^S(J_m^L)e_1\right)$$
$$= \frac{1}{\beta_{j+1}}((J_m^L - \alpha_j I_m)z_{j-1} - \beta_j z_{j-2}).$$

and with
$$\hat{z}_j := s_0'(J_m^L)z_j$$
$$= \frac{1}{\beta_{j+1}}((J_m^L - \alpha_j I_m)\hat{z}_{j-1} - \beta_j \hat{z}_{j-2})$$

the computations in (5.3.7) reduce to

$$\langle t\psi_j^S, \psi_j^S \rangle_S = z_j^T J_m^L \hat{z}_j \quad \text{and} \quad \langle \psi_j^S, \psi_j^S \rangle_S = z_j^T \hat{z}_j.$$

Finally, a proper combination of these formulae leads to a robust implementation of the Stieltjes procedure for the generation of the orthonormal polynomials ψ_k^S.

6 Parameter Free Methods

In this chapter we present a survey of so-called *parameter free polynomial iteration method*. These are schemes

$$r_n = p_n(A)r_0, \quad p_n(0) = 1$$

where the residual polynomial p_n is constructed during the iteration process based on certain orthogonality relations. The methods require as input only the system matrix A, the right hand side f, and a starting guess x_0. The name "parameter free" is motivated by the fact that no further information (or parameter) is necessary to start the process. Incidentally, Householder [83] calls these schemes *methods of projection*.

We will consider four different classes. The first three classes are based on the residual polynomials ψ_n^{OR}, ψ_n^{MR}, and ψ_n^{ME}, respectively. The fourth class deals with two variants of the normal equations.

All classes have in common that their orthogonality relations are more or less connected to the inner product induced by the residual (cf. 1.1.13) and the eigenvalues λ_j of A

$$\langle p, q \rangle_{\text{GAL}} = r_0^T p(A) q(A) r_0 = \sum_{j=1}^{L} \sigma_j^2 p(\lambda_j) q(\lambda_j).$$

We remark that one may find a collection of parameter free methods, from a different point of view, also in Ashby, Manteuffel, and Saylor [6], and in Gutknecht [75], respectively.

6.1 Overview

In this section, we briefly introduce the various classes. This time, we state the variational properties of the schemes in terms of vector-norms and of Krylov subspaces.

The OR Approach

This approach is based upon the residual polynomials ψ_n^{OR}, orthogonal with respect to $\langle \cdot, \cdot \rangle_{\text{GAL}}$ (cf. Example 2.4.8)

$$\langle \psi_k^{\text{OR}}, \psi_l^{\text{OR}} \rangle_{\text{GAL}} = 0, \quad l < k.$$

The iterates $x_n^{\text{OR}} \in x_0 + \mathcal{K}_n(A, r_0)$ –if existing– are uniquely defined by

$$\|x_* - x_n^{\text{OR}}\|_A = \min\{\|x_* - x\|_A : x \in x_0 + \mathcal{K}_n(A, r_0)\}.$$

Alternatively, they are characterized by the Galerkin condition

$$r_n^{\text{OR}} \perp \mathcal{K}_n(A, r_0), \quad r_n^{\text{OR}} = f - Ax_n^{\text{OR}}.$$

The MR Approach

Here, the method is defined by the minimal residual polynomials ψ_n^{MR}, the scaled kernel polynomials with respect to $\langle \cdot, \cdot \rangle_{\text{GAL}}$ (cf. Example 2.5.2)

$$\langle \psi_k^{\text{MR}}, t\psi_l^{\text{MR}} \rangle_{\text{GAL}} = 0, \quad l < k.$$

The iterates $x_n^{\text{MR}} \in x_0 + \mathcal{K}_n(A, r_0)$ are characterized by the minimal residual property

$$\|f - Ax_n^{\text{MR}}\|_2 = \min\{\|f - Ax\|_2 : x \in x_0 + \mathcal{K}_n(A, r_0)\},$$

or, equivalently, by the Petrov Galerkin condition

$$r_n^{\text{MR}} \perp A\mathcal{K}_n(A, r_0), \quad r_n^{\text{MR}} = f - Ax_n^{\text{MR}}.$$

The ME Approach

This time, the Hermite kernel polynomials ψ_n^{ME} enter into the picture (cf. Example 2.6.2). The corresponding iterates $x_n^{\text{ME}} \in x_0 + A\mathcal{K}_n(A, r_0)$ are defined by the minimal Euclidian error property

$$\|x_* - x_n^{\text{ME}}\|_2 = \min\{\|x_* - x\|_2 : x \in x_0 + A\mathcal{K}_n(A, r_0)\}.$$

Again, the residuals fulfill a Galerkin condition

$$r_n^{\text{ME}} \perp \mathcal{K}_n(A, r_0), \quad r_n^{\text{ME}} = f - Ax_n^{\text{ME}}.$$

The next approach is only of academic interest.

6.1 Overview

A "Non-Feasible" Approach

In view of the ME approach it is natural to ask, whether there exists an effective way for the computation of the iterates x_n^{MtE} with minimal Euclidian error with respect to the "true" Krylov subspace $x_0 + \mathcal{K}_n(A, r_0)$, i.e.,

$$\|x_* - x_n^{\text{MtE}}\|_2 = \min\left\{\|x_* - x\|_2 : x \in x_0 + \mathcal{K}_n(A, r_0)\right\}.$$

It turns out that an economic computation of x_n^{MtE} involves the knowledge of the solution x_* of the given system, which is not feasible. On the hand, however, it is interesting to compare the lower bound $\|x_* - x_n^{\text{MtE}}\|_2$ to the corresponding quantity of the other methods.

The Normal Equations Approach

A quite popular way of iteratively solving indefinite systems is the application of the classical CG method (cf. Section 6.2) onto the normal equations and thereby removing obstacles hidden in the indefinite nature of the system. In order to compare this "universal approach" with the "real" indefinite solvers we will as well briefly outline implementation details for this method.

The normal equations come in two versions

$$A^2 x = Af,$$

and

$$A^2 y = f, \quad Ay = x.$$

Applying CG to the first version leads to iterates $x_n^{\text{LSQR}} \in x_0 + A\mathcal{K}_n(A^2, r_0)$ which are characterized by the minimal residual property

$$\|f - Ax_n^{\text{LSQR}}\|_2 = \min\left\{\|f - Ax\|_2 : x \in x_0 + A\mathcal{K}_n(A^2, r_0)\right\}.$$

The residual polynomials $\psi_n^{\text{LSQR}} \in \Pi_n$ in

$$r_n^{\text{LSQR}} = f - Ax_n^{\text{LSQR}} = \psi_n^{\text{LSQR}}(A^2)r_0, \quad r_0 = f - Ax_0,$$

are uniquely determined by the property

$$r_0^T \psi_k^{\text{LSQR}}(A^2) A^2 \psi_l^{\text{LSQR}}(A^2) r_0 = 0, \quad l < k.$$

The superscript LSQR indicates that we will discuss the LSQR implementation of Paige and Saunders [96]. This is a particular stable implementation of the Galerkin scheme based on a *Golub/Kahan bidiagonalization* of the system matrix A (cf. Section 6.8).

The latter version of the normal equations in conjunction with the CG method leads to iterates $x_n^{\text{CRAIG}} \in x_0 + A\mathcal{K}_n(A^2, r_0)$ defined by the minimal error property

$$\|x_* - x_n^{\text{CRAIG}}\|_2 = \min\{\|x_* - x\|_2 : x \in x_0 + A\mathcal{K}_n(A^2, r_0)\}.$$

This time the residual polynomials $\psi_n^{\text{CRAIG}} \in \Pi_n$ in

$$\varepsilon_n^{\text{CRAIG}} = x_* - x_n^{\text{CRAIG}} = \psi_n^{\text{CRAIG}}(A^2)\varepsilon_0, \quad \varepsilon_0 = x_* - x_0,$$

are defined by the orthogonality property

$$\varepsilon_0^{\text{T}} \psi_k^{\text{CRAIG}}(A^2) A^2 \psi_l^{\text{CRAIG}}(A^2) \varepsilon_0 = 0, \quad l < k.$$

This method is due to Craig [24]. We will, however, discuss an implementation of Paige [94] (cf. Section 6.8).

It is the goal of this chapter to discuss implementation details of the various schemes. In view of the discussion above one is tempted to present the schemes in an enumeration according to the introduced classes. We will not follow these lines. Instead we will sort the algorithm according to their main implementation characteristics.

We first discuss two implementations based on the original three-term recurrence of the residual polynomials. This will lead us to the classical CG implementation of the OR scheme, and to the classical CR implementation of the MR scheme, respectively.

Next, we discuss two implementations which are essentially based on a particular basis of the underlying Krylov subspace. Namely, the one provided by the monic polynomials with respect to $\langle \cdot, \cdot t^2 \rangle$. Therefore we refer to these implementations as "based on the monic basis". We obtain the STOD implementation of the ME scheme, and the MCR implementation of the MR scheme, respectively.

The next pair of algorithms is (directly) based on the Lanczos method. Here, we discuss the SYMMLQ implementation of the ME scheme, and the MINRES implementation of the MR scheme, respectively. It is worth noticing that both implementations generate as well the Galerkin iterates, if they exist.

Finally, we investigate two implementations of normal equations solvers. Both are based on the Golub/Kahan bidiagonalization procedure. We discuss the CRAIG implementation of the ME scheme, and the LSQR implementation of the MR scheme, respectively.

6.2 Implementations Based on Three - Term Recurrences

In this section we will discuss implementation details of polynomial iteration methods

$$(6.2.1) \qquad r_n = \psi_n^{\mathrm{OR}}(A) r_0.$$

based on orthogonal residual polynomials (cf. Section 2.4)

$$\psi_{-1}^{\mathrm{OR}}(t) := 0, \quad \psi_0^{\mathrm{OR}}(t) = 1,$$
$$\gamma_j \psi_j^{\mathrm{OR}}(t) = (t + (\gamma_j + \beta_j)) \psi_{j-1}^{\mathrm{OR}}(t) - \beta_j \psi_{j-2}^{\mathrm{OR}}(t), \quad j \geq 1.$$

We first work out a general framework for such algorithms. That is, we assume that the three-term recurrence coefficients β_j and α_j are given. Based on this information we then generate a "prototype algorithm" for schemes defined by orthogonal residual polynomials.

Basic Algorithm

To devise a scheme for the computation of the iterates x_n, implicitly defined by (6.2.1), note that the residuals inherit the three-term recurrence property

$$\begin{aligned} A r_{n-1} &= A \psi_{n-1}^{\mathrm{OR}}(A) r_0 \\ &= \gamma_n \psi_n^{\mathrm{OR}}(A) r_0 - (\gamma_n + \beta_n) \psi_{n-1}^{\mathrm{OR}}(A) r_0 + \beta_n \psi_{n-2}^{\mathrm{OR}}(A) r_0 \\ &= \gamma_n r_n - (\gamma_n + \beta_n) r_{n-1} + \beta_n r_{n-2}. \end{aligned}$$

Hence, for the residual increment we obtain

$$\Delta r_{n-1} := r_n - r_{n-1} = \frac{1}{\gamma_n} (\beta_n \Delta r_{n-2} + A r_{n-1}),$$

which translates for the iterates

$$\Delta x_{n-1} := x_n - x_{n-1} = -A^{-1} \Delta r_{n-1} = \frac{1}{\gamma_n} (\beta_n \Delta x_{n-2} - r_{n-1}).$$

To update the iterates we obtain

$$x_n = x_{n-1} + \frac{1}{\gamma_n} w_{n-1},$$

where

(6.2.2) $$w_{n-1} = \frac{\beta_n}{\gamma_{n-1}} w_{n-2} - r_{n-1}.$$

The update formula for the residuals involves as well the *search direction* w_{n-1}

(6.2.3) $$r_n = r_{n-1} - \frac{1}{\gamma_n} A w_{n-1}$$

Altogether, we arrive at the following scheme, the "prototype" of polynomial iteration methods based on orthogonal residual polynomials.

Algorithm 6.2.1 *Let $\{\psi_n^{OR}, (\beta_n, \gamma_n); \langle \cdot, \cdot \rangle\}$ (cf. (2.1.13)) and x_0 be given. The iterates x_n and residuals r_n, defined by $r_n = \psi_n^{OR}(A) r_0$, may be computed by the following recursive procedure.*

Set
$$w_{-1} := 0, \quad r_0 = f - A x_0.$$

For $n = 1, 2, \ldots$ compute
$$w_{n-1} = \frac{\beta_n}{\gamma_{n-1}} w_{n-2} - r_{n-1},$$
$$x_n = x_{n-1} + \frac{1}{\gamma_n} w_{n-1},$$
$$r_n = r_{n-1} - \frac{1}{\gamma_n} A w_{n-1}.$$

Given the three-term recurrence coefficients the Algorithm above is quite cheap. It requires per loop only one matrix-vector multiplication $A w_{n-1}$ and $6N$ additional flops. Note that $A w_{n-1}$ may be updated in terms $A r_{n-1}$

$$A w_{n-1} = \frac{\beta_n}{\gamma_{n-1}} A w_{n-2} - A r_{n-1}.$$

This follows from (6.2.2). In addition, (6.2.2) implies that

(6.2.4) $$w_{n-1} \in \text{span}\{r_0, r_1, \ldots, r_{n-1}\},$$

6.2 Implementations Based on Three - Term Recurrences

whereas (6.2.3) yields

(6.2.5) $$Aw_{n-1} \in \text{span}\{r_{n-1}, r_n\}.$$

The CG Approach

In this section we work out an implementation for the polynomial iteration method
$$r_n^{\text{CG}} = \psi_n^{\text{OR}}(A)r_0.$$
defined by the residual polynomials ψ_n^{OR}, orthogonal with respect to $\langle \cdot, \cdot \rangle_{\text{GAL}}$ (cf. Example 2.4.8). One way to generate these polynomials is provided by the Stieltjes-like Algorithm 2.4.2. We will now show that a translation of this scheme into the language of polynomial iteration methods, together with Algorithm 6.2.1, leads to the standard implementation of the celebrated *conjugate gradient algorithm* (see, e.g., Golub and Van Loan [71, §10.2]). Therefore, all participating quantities are equipped with the superscript CG.

To devise the implementation we only have to generate efficient formulae for the computation of the three-term recurrence coefficients β_n and γ_n. The calculation of β_n/γ_{n-1} is straightforward. Working from Algorithm 2.4.2, we have

(6.2.6) $$\nu_n^{\text{CG}} := \frac{\beta_n}{\gamma_{n-1}} = \frac{\langle \psi_{n-1}^{\text{OR}}, \psi_{n-1}^{\text{OR}} \rangle_{\text{GAL}}}{\langle \psi_{n-2}^{\text{OR}}, \psi_{n-2}^{\text{OR}} \rangle_{\text{GAL}}} = \frac{(r_{n-1}^{\text{CG}})^{\text{T}} r_{n-1}^{\text{CG}}}{(r_{n-2}^{\text{CG}})^{\text{T}} r_{n-2}^{\text{CG}}}.$$

To compute $1/\gamma_n$, we could use the expression $\gamma_n = -(\alpha_n + \beta_n)$. We will instead use the "direct representation" (2.1.16)

(6.2.7) $$\eta_n^{\text{CG}} := \frac{1}{\gamma_n} = \frac{\langle \psi_n^{\text{OR}}, \psi_n^{\text{OR}} \rangle_{\text{GAL}}}{\langle t\psi_{n-1}^{\text{OR}}, \psi_n^{\text{OR}} \rangle_{\text{GAL}}} = \frac{(r_n^{\text{CG}})^{\text{T}} r_n^{\text{CG}}}{(Ar_{n-1}^{\text{CG}})^{\text{T}} r_n^{\text{CG}}}.$$

This formula, however, involves r_n^{CG} which is, in view of (6.2.3), prohibitive. To overcome this problem, observe that by (6.2.4), (6.2.5) and the Galerkin condition (2.4.7) the search direction (6.2.2) are mutually A-orthogonal

$$(w_n)^{\text{T}} A w_k = 0, \quad k < n.$$

With this orthogonality relation and (6.2.2) we obtain

$$(Ar_{n-1}^{\text{CG}})^{\text{T}} r_n^{\text{CG}} = -\frac{\beta_{n+1}}{\gamma_n} w_{n-1}^{\text{T}} A w_{n-1} = -\frac{(r_n^{\text{CG}})^{\text{T}} r_n^{\text{CG}}}{(r_{n-1}^{\text{CG}})^{\text{T}} r_{n-1}^{\text{CG}}} w_{n-1}^{\text{T}} A w_{n-1},$$

which yields in conjunction with (6.2.7) the computable expression

$$\eta_n^{CG} = -\frac{(r_{n-1}^{CG})^T r_{n-1}^{CG}}{w_{n-1}^T A w_{n-1}}. \tag{6.2.8}$$

Algorithm 6.2.1 together with (6.2.6) and (6.2.8) furnishes the CG algorithm.

Theorem 6.2.2 *The nth CG iterate and its residual are given by*

$$x_n^{CG} = x_{n-1}^{CG} + \eta_n^{CG} w_{n-1}, \quad x_0^{CG} = x_0,$$
$$r_n^{CG} = r_{n-1}^{CG} - \eta_n^{CG} A w_{n-1}, \quad r_0^{CG} = r_0,$$

where

$$\eta_n^{CG} = -\frac{(r_{n-1}^{CG})^T r_{n-1}^{CG}}{w_{n-1}^T A w_{n-1}},$$

and

$$w_{n-1} = \frac{(r_{n-1}^{CG})^T r_{n-1}^{CG}}{(r_{n-2}^{CG})^T r_{n-2}^{CG}} w_{n-2} - r_{n-1}^{CG}, \quad w_{-1} = 0, \quad w_0 = -r_0.$$

Let us say a few words to the case where the iterate x_n^{CG} does not exist. According to Lemma 2.4.1 this is precisely the case if $\gamma_n = 0$. Hence, the degenerate case is characterized by the fact that the "stepsize" η_n^{CG} is ∞. In the following MATLAB implementation of the CG method we do not take care of the degenerate case. Anyway, if the system under consideration happens to be positive definite, i.e., all zeros of ψ_n are positive, the scheme cannot break down. In Section 6.3 we will apply the implementation as well to indefinite systems and comment on some peculiarities.

MATLAB Implementation of CG

Notice that the norm of the residual can be conveniently computed by taking the square root of the dot product **rr**, which has to be computed anyway. As a safeguard, however, one should compare this quantity, towards the end of the iteration, with the norm of the "true residual" $\|f - A x_n^{CG}\|_2$.

6.2 Implementations Based on Three - Term Recurrences

```
function [xCG,norm_rCG]=CG(A,f,x0,n_max,tol)
%
% computes the solution of Ax=f using the CG scheme.
%
% startvector:                            x0
% maximal number of iterations:          n_max
% desired accuracy of the residual: tol
%
% uses: matvec
%

%%initialize
   n=1; N=length(x0);
   xCG=x0; rCG=f-feval(matvec,A,xCG);
   rr=rCG'*rCG; rr_old=1;
   norm_rCG=sqrt(rr); norm_r0=norm_rCG;
   w=zeros(N,1);

%%iterate
   while (n < n_max+1) &  (norm_rCG/norm_r0 > tol)
      n=n+1;

      nu =rr/rr_old;
      w  =nu*w-rCG;
      Aw =matvec(A,w);
      eta=-rr/(w'*Aw);

   %%update
      xCG=xCG+eta*w;
      rCG=rCG-eta*Aw;

      rr_old=rr; rr=rCG'*rCG;

   %%compute norm
      norm_rCG=sqrt(rr);

   end %while

return;
```

Work per iteration and storage requirements of CG:

- Matrix-vector products: `matvec(A,w)`   1
- dot products: `rCG'*rCG`, `w'*Aw`   2
- additional flops: `nu*w-rCG`, `xCG+eta*w`,
                    `rCG-eta*Aw`   6N
- vectors to be stored: `w, rCG, Aw, xCG`   4

The CR Approach

Here, we discuss a particular implementation of the method

$$r_n^{\text{CR}} = \psi_n^{\text{MR}}(A) r_0$$

defined by the scaled kernel polynomial ψ_n^{MR} (cf. (2.5.5) and Example 2.5.2). So, we are dealing with a conjugate residual (CR) method

(6.2.9) $$(r_n^{\text{CR}})^{\text{T}} A r_k^{\text{CR}} = 0, \quad k < n.$$

As in the "CG case", we generate the residual polynomials by means of their three-term recurrence relation as written down in Algorithm 2.5.5. It should come as no surprise that this time the classical CR implementation results (see *e.g.*, Stiefel [115]).

In view of Algorithm 2.5.5 we obtain

$$\nu_n^{\text{CR}} := \frac{\beta_n}{\gamma_{n-1}} = \frac{\langle \psi_{n-1}^{\text{MR}}, t\psi_{n-1}^{\text{MR}} \rangle_{\text{GAL}}}{\langle \psi_{n-2}^{\text{MR}}, t\psi_{n-2}^{\text{MR}} \rangle_{\text{GAL}}} = \frac{(r_{n-1}^{\text{CR}})^{\text{T}} A r_{n-1}^{\text{CR}}}{(r_{n-2}^{\text{CR}})^{\text{T}} A r_{n-2}^{\text{CR}}},$$

and

(6.2.10) $$\eta_n^{\text{CR}} := \frac{1}{\gamma_n} = \frac{\langle \psi_n^{\text{MR}}, t\psi_n^{\text{MR}} \rangle_{\text{GAL}}}{\langle t\psi_{n-1}^{\text{MR}}, t\psi_n^{\text{MR}} \rangle_{\text{GAL}}} = \frac{(r_n^{\text{CR}})^{\text{T}} A r_n^{\text{CR}}}{(A r_{n-1}^{\text{CR}})^{\text{T}} A r_n^{\text{CR}}}.$$

Again, the formula for the stepsize η_n^{CR} needs some further investigations. However, observe that (6.2.4), (6.2.5) and (6.2.9) imply

$$(A w_{n-1})^{\text{T}} A w_k = 0, \quad k < n - 1.$$

A similar derivation as for η_n^{CG} (cf. (6.2.8)) finally yields the computable expression

$$\eta_n^{\text{CR}} = -\frac{(r_{n-1}^{\text{CR}})^{\text{T}} A r_{n-1}^{\text{CR}}}{(A w_{n-1})^{\text{T}} A w_{n-1}}.$$

6.2 Implementations Based on Three - Term Recurrences

Theorem 6.2.3 *The nth CR iterate and its residual are given by*

$$x_n^{CR} = x_{n-1}^{CR} + \eta_n^{CR} w_{n-1}, \quad x_0^{CR} = x_0,$$
$$r_n^{CR} = r_{n-1}^{CR} - \eta_n^{CR} A w_{n-1}, \quad r_0^{CR} = r_0,$$

where

$$\eta_n^{CR} = -\frac{(r_{n-1}^{CR})^T A r_{n-1}^{CR}}{(Aw_{n-1})^T Aw_{n-1}},$$

and

$$w_{n-1} = \frac{(r_{n-1}^{CR})^T A r_{n-1}^{CR}}{(r_{n-2}^{CR})^T A r_{n-2}^{CR}} w_{n-2} - r_{n-1}^{CR}, \quad w_{-1} = 0, \quad w_0 = -r_0.$$

For indefinite systems the degenerate case $\psi_n(0) = 0$ may occur. This implies, by Lemma 2.5.4(c), that $\langle \psi_{n-1}^{MR}, t\psi_{n-1}^{MR} \rangle_{GAL} = 0$. Consequently, we then have to face $\nu_{n+1}^{CR} = \infty$, a breakdown.

MATLAB Implementation of CR

The operation count for CR looks like

Work per iteration and storage requirements of CR:

- Matrix-vector products: `matvec(A,rCR)`   1
- dot products: `rCR'*Ar, Aw'*Aw`   2
- additional flops: `nu*w-rCR, nu*Aw-Ar,`
  `xCR+eta*w, rCR-eta*Aw`   8N
- vectors to be stored: `Ar, w, rCR, Aw, xCR`   5

The above operation count does not include the work needed for the computation of the norm of the residual `norm(rCR)` which is used as stopping criterion.

```
function [xCR,norm_rCR]=CR(A,f,x0,n_max,tol)
%
% computes the solution of Ax=f using the CR scheme.
%
% startvector:                        x0
% maximal number of iterations:       n_max
% desired accuracy of the residual: tol
%
% uses: matvec
%
%%initialize
    n=1; N=length(x0);
    xCR=x0; rCR=f-matvec(A,xCR);
    norm_rCR=norm(rCR); norm_r0=norm_rCR;
    rAr=1;
    w=zeros(N,1); Aw=zeros(N,1);

%%iterate
    while (n < n_max+1) & (norm_rCR/norm_r0 > tol)
       n=n+1;

       rAr_old=rAr;
       Ar  =matvec(A,rCR); rAr=rCR'*Ar;
       nu  =rAr/rAr_old;
       w   =nu*w-rCR; Aw=nu*Aw-Ar;
       eta =-rAr/(Aw'*Aw);

%%update
       xCR=xCR+eta*w;
       rCR=rCR-eta*Aw;

%%compute norm
       norm_rCR=norm(rCR);

    end %while
return;
```

6.3 CG/CR Applied to Indefinite Systems

Since all zeros of ψ_n are located in the interval $[\lambda_1, \lambda_N]$ (cf. Cor. 5.2.1(c)), the degenerate case $\psi_n(0) = 0$ can not occur for positive definite systems. This situation changes for symmetric indefinite cases. Here a breakdown of CG or CR is possible. On the other hand, from a "generic point of view", the situation $\psi_n(0) = 0$ is unlikely to take place and consequently CG and CR may well succeed in producing satisfactory approximations for these linear systems. In this section we illustrate the behavior of CG and CR, respectively, for indefinite matrices. To this end, we will discuss two (academic) examples, each from one side of the spectrum, i.e., for the

6.3 CG/CR Applied to Indefinite Systems

first experiment both CG and CR behave nicely, whereas for the second experiment the performance of these methods is rather bad.

Example 6.3.1 We start with an experiment taken from Paige, Parlett and van der Vorst [93, Sec. 4]. Here A is a diagonal matrix of dimension $N = 100$

(6.3.1) $\qquad A = \text{diag}(-9, -7, -5, -3, -1, 1, 3, \ldots, 189).$

Note that A has 5 negative eigenvalues. The right hand side f was chosen such that $x_* = (1, 1, \ldots, 1)^T$ is the solution. A random vector with components uniformly distributed in $[-1, 1]$ was selected as initial approximation x_0.

To demonstrate the effectiveness of the devised schemes we run them until the very end, i.e., the iterations were halted once the relative residual was at about machine precision

$$\frac{\|r_n\|_2}{\|r_0\|_2} \leq 10^{-15}.$$

As is apparent from Figure 3.4.6, the CG-scheme produces precisely 5 peaks in early stages of the iteration. These are explained as follows (compare [93, Sec. 4]). On the way to their final destination the Ritz values, associated with the 5 negative eigenvalues, have to cross over the origin. Here it may happen that they come close to the origin and hence the corresponding error may expected to be large. To justify this, observe that by (1.1.13)

$$\|r_n\|_2^2 = \sum_{j=1}^{N} \sigma_j^2 (\psi_n^{\text{OR}}(\lambda_j))^2$$

with (cf. (2.4.3))

$$|\psi_n^{\text{OR}}(\lambda_j)| = \prod_{k=1}^{n} |1 - \lambda_j/\theta_k^{(n)}|.$$

So, if some Ritz value $\theta_k^{(n)}$ is very small relative to the eigenvalue λ_j, the error can be very high indeed. The peaks towards the end of the iteration are due to roundoff errors.

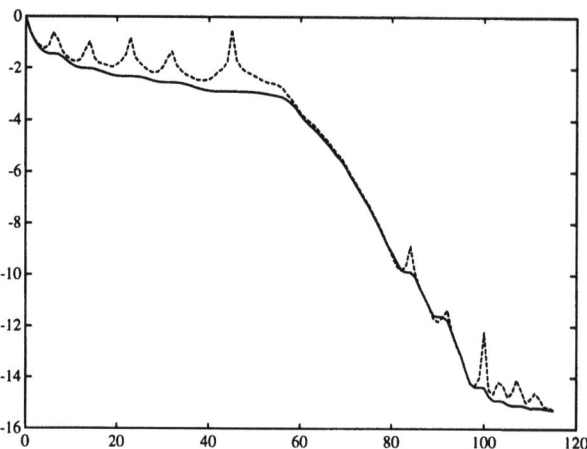

Figure 6.3.2 Residual norms versus the number of iterations for the matrix (6.3.1). Solid curve: $\log_{10} \|f - Ax_n^{\text{CR}}\|_2/\|r_0\|_2$; dashed curve: $\log_{10} \|f - Ax_n^{\text{CG}}\|_2/\|r_0\|_2$.

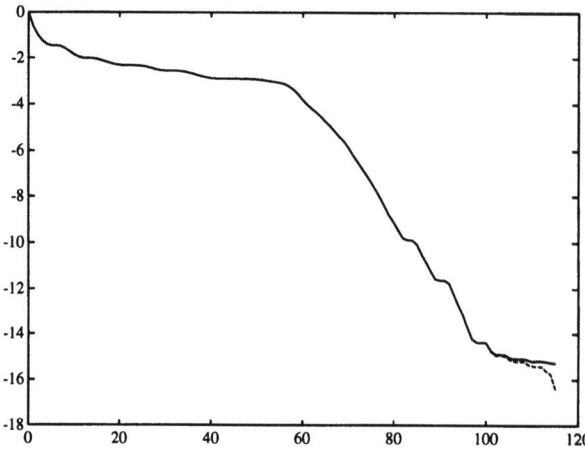

Figure 6.3.3 Updated and true CR residual norms versus the number of iterations for the matrix (6.3.1). Solid curve: $\log_{10} \|f - Ax_n^{\text{CR}}\|_2/\|r_0\|_2$; dashed curve: $\log_{10} \|r_n^{\text{CR}}\|_2/\|r_0\|_2$.

6.3 CG/CR Applied to Indefinite Systems

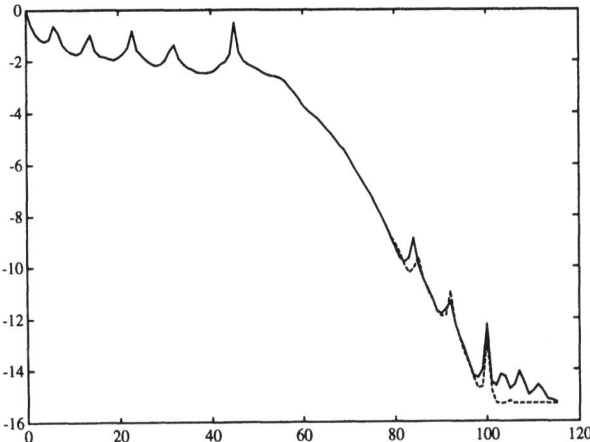

Figure 6.3.4 Residual norms versus the number of iterations for the matrix (6.3.1). Solid curve: $\log_{10} \|f - Ax_n^{CG}\|_2/\|r_0\|_2$; dashed curve: $\log_{10} \|f - Ax_n^{OR}\|_2/\|r_0\|_2$.

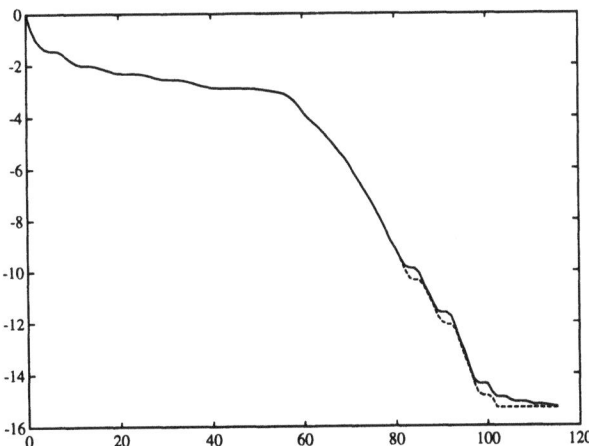

Figure 6.3.5 Residual norms versus the number of iterations for the matrix (6.3.1). Solid curve: $\log_{10} \|f - Ax_n^{CR}\|_2/\|r_0\|_2$; dashed curve: $\log_{10} \|f - Ax_n^{MR}\|_2/\|r_0\|_2$.

Next (see Figure 6.3.3), we compare the updated residual r_n^{CR} with the true residual $f - Ax_n^{CR}$. In the absence of roundoff errors they should be the same and thus a possible divergence of these quantities would indicate an unstable implementation. Note that there is no significant difference between the two curves. The same qualitative behavior can be observed in the corresponding CG case.

It is interesting to compare the error curves in the Figure 6.3.2 with the one obtained from stable, though more expensive, implementations of CG and CR, respectively. These implementations, more precisely SYMMLQ and MINRES, will be discussed in detail in the next subsections.

It comes at a little surprise (compare Figure 6.3.4 and Figure 6.3.5) that for this example the CG and CR implementations turn out to be very stable. Except for the very last stages of the iteration the respective error curves are visually indistinguishable.

We mention that although the convergence curve of the CR scheme looks rather smooth, this not necessarily implies that the current implementation is stable. This will be apparent from the next example.

Example 6.3.6 For this experiment we artificially generated a small Ritz value. More precisely, we shifted the matrix A (cf. (6.3.1)) from the previous example

$$(6.3.2) \qquad A = A - 0.967156 * I_N,$$

but kept the very same starting residual r_0. Thus, in view of Cor. 5.2.2(a), we obtain as well a shifted set of Ritz values $\theta_j^{(n)} = \theta_j^{(n)} - 0.967156$. In particular the smallest computed positive Ritz value of J_{22} becomes $\theta_3^{(22)} = 8.48\ldots * 10^{-7}$. We stress that the matrix A itself is well conditioned, since all eigenvalues are located in $[-10, -1.9] \cup [0.03, 188.2]$. Moreover, A still has only 5 negative eigenvalues.

We start by illustrating that this time the norm of the updated residuals r_n and the norm of the true residuals $f - Ax_n$ diverge from each other. It is apparent, from the next two figures, that the updated residuals converge to full machine precision, whereas the relative norm of the true residuals stays constant at about 10^{-11}. In addition note that the CG scheme has for $j = 22$ a peak of height about 10^{+5}. These two observations suggest the following heuristic for the true residuals

$$(6.3.3) \qquad \min_n \frac{\|f - Ax_n^{CG}\|_2}{\|r_0\|_2} \approx \text{EPS} + \max_n \frac{\|f - Ax_n^{CG}\|_2}{\|r_0\|_2},$$

6.3 CG/CR Applied to Indefinite Systems

where EPS$\approx 10^{-16}$ denotes the machine precision. Actually, it has been proven by Sleijpen, van der Vorst and Fokkema [113] that

$$\left| \|r_n^{CG}\|_2 - \|f - Ax_n^{CG}\|_2 \right| \leq n \cdot c \cdot \text{EPS} \max_{j \leq n} \|r_n^{CG}\|_2,$$

where c is a "modest" multiple of the condition number of A. We conclude that iterations with large residuals may ultimately lead to inaccurate results. Also, notice that (6.3.3) is as well valid for the first experiment.

It is interesting to see how the convergence curves in Figure 6.3.7 and Figure 6.3.8 relate to the ones produced by stable implementations. We clearly see that the occurrence of small Ritz values may heavily affect the convergence behavior of both the CG scheme and the CR scheme. In Example 6.3.1 we explained the peaks in the convergence curve of the CG iteration with the fact that Ritz values have to converge to negative eigenvalues. In the present example the CG curve is much more erratic, whereas the curve obtained from the stable implementation SYMMLQ again has 5 (main) peaks. Moreover, after this startup phase the process behaves like a process for positive definite systems.

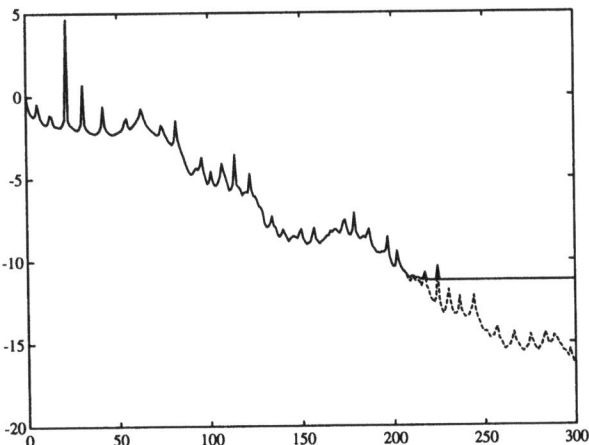

Figure 6.3.7 Updated and true CG residual norms versus the number of iterations for the matrix (6.3.2). Solid curve: $\log_{10} \|f - Ax_n^{CG}\|_2/\|r_0\|_2$; dashed curve: $\log_{10} \|r_n^{CG}\|_2/\|r_0\|_2$.

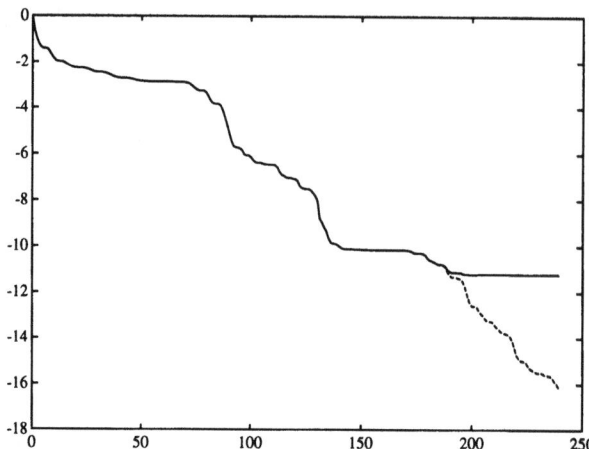

Figure 6.3.8 Updated and true CR residual norms versus the number of iterations for the matrix (6.3.2). Solid curve: $\log_{10} \|f - Ax_n^{CR}\|_2 / \|r_0\|_2$; dashed curve: $\log_{10} \|r_n^{CR}\|_2 / \|r_0\|_2$.

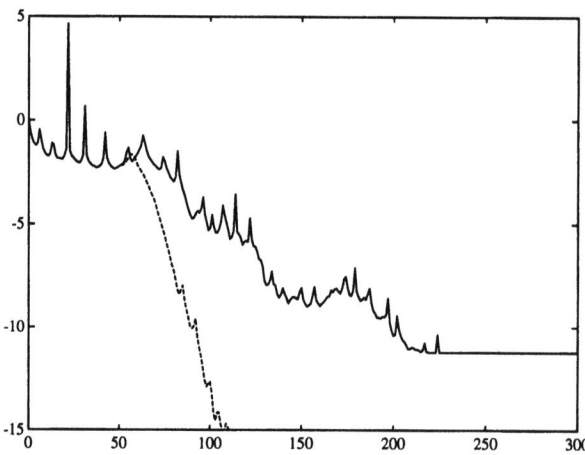

Figure 6.3.9 Residual norms versus the number of iterations for the matrix (6.3.2). Solid curve: $\log_{10} \|f - Ax_n^{CG}\|_2 / \|r_0\|_2$; dashed curve: $\log_{10} \|f - Ax_n^{OR}\|_2 / \|r_0\|_2$.

6.4 Implementations Based on the Monic Basis

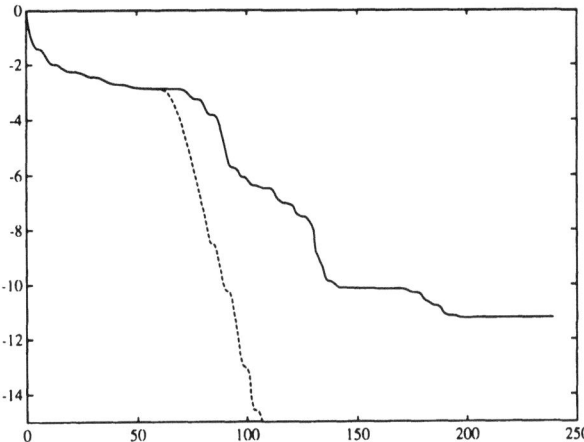

Figure 6.3.10 Residual norms versus the number of iterations for the matrix (6.3.2). Solid curve: $\log_{10}\|f - Ax_n^{\text{CR}}\|_2/\|r_0\|_2$; dashed curve: $\log_{10}\|f - Ax_n^{\text{MR}}\|_2/\|r_0\|_2$.

The last two figures leave us with the question: how much more expensive are the stable implementations of CG and CR compared to the "classical ones"? This question will be answered in the next sections.

6.4 Implementations Based on the Monic Basis

In this section we will discuss two particular implementations of polynomial iteration methods. Again, the underlying inner product is given by $\langle \cdot, \cdot \rangle_{\text{GAL}}$. The first implementation generates the residual polynomials

$$(6.4.1) \qquad r_n^{\text{STOD}} = \psi_n^{\text{ME}}(A) r_0$$

with the help of Algorithm 2.6.6, whereas the second one

$$(6.4.2) \qquad r_n^{\text{MCR}} = \psi_n^{\text{MR}}(A) r_0$$

is based on Algorithm 2.5.6.

Recall that both algorithms compute as well the monic orthogonal polynomials ψ_n^{MO} associated with $\langle \cdot, \cdot t^2 \rangle_{\text{GAL}}$. As demonstrated in Section 4.2, these polynomials can be used to generate a basis for the corresponding Krylov subspace (cf. (4.2.1))

$$(6.4.3) \qquad \text{span}\{v_1^{\text{MO}}, v_2^{\text{MO}}, \ldots, v_n^{\text{MO}}\} = \mathcal{K}_n(A, r_0), \quad v_j^{\text{MO}} := \psi_{j-1}^{\text{MO}}(A) r_0.$$

The STOD Approach

Here, we present a stable implementation of the ME scheme (6.4.1) defined by the scaled Hermite kernel polynomials ψ_n^{ME} (cf. (2.6.11) and Example 2.6.2).

Before we discuss implementation details, some historical remarks are in order. The "ME business" started with a paper by Fridman [53]. His method, however, is numerically unstable. Thirteen years later, Fletcher [46] devised a scheme (the OD algorithm) which turns out to be equivalent to Fridman's method, as was shown by Freund [47]. Later on, Stoer and Freund [119] (see also [47]) presented a stable variant of the OD scheme, the STOD algorithm.

It is precisely the last algorithm we are about to discuss. In view of Algorithm 2.6.6 and (6.4.3), it remains to devise update formulae for r_n^{STOD} and x_n^{STOD}, respectively. We have

$$\rho_{n-1}^{\text{STOD}} = \frac{\langle \psi_{n-1}^{\text{ME}}, \psi_{n-1}^{\text{MO}} \rangle_{\text{GAL}}}{\langle \psi_{n-1}^{\text{MO}}, t^2 \psi_{n-1}^{\text{MO}} \rangle_{\text{GAL}}} = \frac{(r_{n-1}^{\text{STOD}})^{\text{T}} v_n^{\text{MO}}}{(Av_n^{\text{MO}})^{\text{T}} Av_n^{\text{MO}}},$$

and

$$\begin{aligned} r_n^{\text{STOD}} &= \psi_n^{\text{ME}}(A) r_0 \\ &= \psi_{n-1}^{\text{ME}}(A) r_0 - \rho_{n-1}^{\text{STOD}} A^2 \psi_{n-1}^{\text{MO}}(A) r_0 \\ &= r_{n-1}^{\text{STOD}} - \rho_{n-1}^{\text{STOD}} A^2 v_n^{\text{MO}}. \end{aligned}$$

Note that the last equation immediately implies

$$x_n^{\text{STOD}} = x_{n-1}^{\text{STOD}} + \rho_{n-1}^{\text{STOD}} Av_n^{\text{MO}}.$$

Altogether we obtain the following theorem.

> **Theorem 6.4.1** *Let v_n^{MO} (cf. (4.2.1)) denote the basis vectors generated by Algorithm 4.2.1 starting with $v_1^{\text{MO}} = r_0$.*
>
> *The nth STOD iterate and its residual are given by*
>
> $$x_n^{\text{STOD}} = x_{n-1}^{\text{STOD}} + \rho_{n-1}^{\text{STOD}} Av_n^{\text{MO}}, \quad x_0^{\text{STOD}} = x_0,$$
>
> $$r_n^{\text{STOD}} = r_{n-1}^{\text{STOD}} - \rho_{n-1}^{\text{STOD}} A^2 v_n^{\text{MO}}, \quad r_0^{\text{STOD}} = r_0,$$
>
> *where*
>
> $$\rho_{n-1}^{\text{STOD}} = \frac{(r_{n-1}^{\text{STOD}})^{\text{T}} v_n^{\text{MO}}}{(Av_n^{\text{MO}})^{\text{T}} Av_n^{\text{MO}}}.$$

6.4 Implementations Based on the Monic Basis

MATLAB Implementation of STOD

```
function [xOD,norm_rOD]=STOD(A,f,x0,n_max,tol)
%
% computes the solution of Ax=f using the STOD scheme.
%
% startvector:                        x0
% maximal number of iterations:       n_max
% desired accuracy of the residual:   tol
%
% uses: matvec
%

%%initialize
   n=1; N=length(x0);
   v=f-matvec(A,x0); v_old=zeros(N,1); beta=0;
   Av=matvec(A,v);   Av_old=zeros(N,1);
   xOD=x0; rOD=v;
   norm_rOD=norm(v); norm_r0=norm_rOD;

   while (n<n_max+1) & (norm_rOD/norm_r0>tol)
       n=n+1;
       AAv=matvec(A,Av); AvAv=Av'*Av;
       rho=rOD'*v/AvAv;

   %%update
       xOD=xOD+rho*Av;
       rOD=rOD-rho*AAv;

   %%compute norm
       norm_rOD=norm(rOD);

   %%monic polynomial
       alpha=Av'*AAv/AvAv;
       if n > 2, beta=AvAv/AvAv_old; end;

       v_oold=v_old; v_old=v;
       v =Av-alpha*v_old-beta*v_oold;
       Av_oold=Av_old; Av_old=Av;
       Av=AAv-alpha*Av_old-beta*Av_oold;
       AvAv_old=AvAv;

   end %while
return;
```

Note that `v_oold` and `Av_oold` can share one storage place. One could actually save one more storage place, in the "monic polynomial part", if there would be a `swap` function in MATLAB, i.e., a subroutine which exchanges the contents of two storage places by just exchanging their pointers. Furthermore, the operation count does not include the work needed for the evaluation of `norm(rOD)`.

> Work per iteration and storage requirements of STOD:
>
> - Matrix-vector products: matvec(A,Av) 1
> - dot products: Av'*Av, rOD'*v, Av'*AAv 3
> - additional flops: xOD+rho*Av, rOD-rho*AAv,
>   Av-alpha*v_old-beta*v_oold,
>   AAv-alpha*Av_old-beta*Av_oold 12N
> - vectors to be stored: rOD, xOD, AAv, v, v_old,
>   v_oold, Av, Av_old 8

The MCR Approach

In this section we devise a stable implementation of the MR scheme (6.4.2) defined by the scaled kernel polynomial ψ_n^{MR} (cf. (2.5.5) and Example 2.5.2). This particular implementation goes back to Fletcher [46] and Chandra [18] (see, also Chandra, Eisenstat and Schultz [19]).

To apply Algorithm 2.5.6, observe that

$$\rho_{n-1}^{\mathrm{MCR}} = \frac{\langle \psi_{n-1}^{\mathrm{MR}}, t\psi_{n-1}^{\mathrm{MO}} \rangle_{\mathrm{GAL}}}{\langle \psi_{n-1}^{\mathrm{MO}}, t^2 \psi_{n-1}^{\mathrm{MO}} \rangle_{\mathrm{GAL}}} = \frac{(r_{n-1}^{\mathrm{MCR}})^{\mathrm{T}} A v_n^{\mathrm{MO}}}{(Av_n^{\mathrm{MO}})^{\mathrm{T}} A v_n^{\mathrm{MO}}},$$

and

(6.4.4)
$$\begin{aligned} r_n^{\mathrm{MCR}} &= \psi_n^{\mathrm{MR}}(A) r_0 \\ &= \psi_{n-1}^{\mathrm{MR}}(A) r_0 - \rho_{n-1}^{\mathrm{MCR}} A \psi_{n-1}^{\mathrm{MO}}(A) r_0 \\ &= r_{n-1}^{\mathrm{MCR}} - \rho_{n-1}^{\mathrm{MCR}} A v_n^{\mathrm{MO}}. \end{aligned}$$

Finally, the iterates look like

(6.4.5)
$$x_n^{\mathrm{MCR}} = x_{n-1}^{\mathrm{MCR}} + \rho_{n-1}^{\mathrm{MCR}} v_n^{\mathrm{MO}}.$$

Here, we used again the basis vectors v_j^{MO} defined by (6.4.3).

6.4 Implementations Based on the Monic Basis

Theorem 6.4.2 *Let v_n^{MO} (cf. (4.2.1)) denote the basis vectors generated by Algorithm 4.2.1 starting with $v_1^{\text{MO}} = r_0$.*

The nth MCR iterate and its residual are given by

$$x_n^{\text{MCR}} = x_{n-1}^{\text{MCR}} + \rho_{n-1}^{\text{MCR}} v_n^{\text{MO}}, \quad x_0^{\text{MCR}} = x_0,$$

$$r_n^{\text{MCR}} = r_{n-1}^{\text{MCR}} - \rho_{n-1}^{\text{MCR}} A v_n^{\text{MO}}, \quad r_0^{\text{MCR}} = r_0,$$

where

$$\rho_{n-1}^{\text{MCR}} = \frac{(r_{n-1}^{\text{MCR}})^{\text{T}} A v_n^{\text{MO}}}{(A v_n^{\text{MO}})^{\text{T}} A v_n^{\text{MO}}}.$$

Observe the similarities between the STOD scheme (cf. Theorem 6.4.1) and the MCR scheme.

MATLAB Implementation of MCR

Essentially, the same comments as for the STOD implementation are valid. In particular, the operation count does not include the work needed for the evaluation of `norm(rMCR)`.

Work per iteration and storage requirements of MCR:

- Matrix-vector products: `matvec(A,Av)`   1
- dot products: `Av'*Av, rMCR*Av, Av'*AAv`   3
- additional flops: `xMCR+rho*v, rMCR-rho*Av,`
  `Av-alpha*v_old-beta*v_oold,`
  `AAv-alpha*Av_old-beta*Av_oold`   12N
- vectors to be stored: `rMCR, xMCR, AAv, v, v_old,`
  `v_oold, Av, Av_old`   8

```
function [xMCR,norm_rMCR]=MCR(A,f,x0,n_max,tol)
%
% computes the solution of Ax=f using the MCR scheme.
%
% startvector:                             x0
% maximal number of iterations:            n_max
% desired accuracy of the residual: tol
%
% uses: matvec
%
%%initialize
    n=1; N=length(x0);
    v=f-matvec(A,x0); v_old=zeros(N,1); beta=0;
    Av=matvec(A,v);   Av_old=zeros(N,1);
    xMCR=x0; rMCR=v;
    norm_rMCR=norm(v); norm_r0=norm_rMCR;

    while (n<n_max+1) & (norm_rMCR/norm_r0>tol)
        n=n+1;
        AAv=matvec(A,Av);   AvAv=Av'*Av;
        rho=rMCR'*Av/AvAv;

%%update
        xMCR=xMCR+rho*v;
        rMCR=rMCR-rho*Av;

%%compute norm
        norm_rMCR=norm(rMCR);

%%monic polynomial
        alpha=Av'*AAv/AvAv;
        if n > 2, beta=AvAv/AvAv_old; end;

        v_oold=v_old; v_old=v;
        v =Av-alpha*v_old-beta*v_oold;
        Av_oold=Av_old; Av_old=Av;
        Av=AAv-alpha*Av_old-beta*Av_oold;
        AvAv_old=AvAv;

    end %while
return;
```

Modifications

As is apparent from the respective tables, the STOD and MCR implementation, though numerically stable, do not compare favorably in terms operation count and storage requirement to the CG and CR implementation. In this section we will briefly present two less expensive modifications of the STOD and MCR approach, respectively. For details we refer to Chandra [18] (for the MCR scheme) and to Freund [47] (for the STOD scheme).

6.5 Implementations Based on the Lanczos Basis

Let us start with the MCR method. Here, in each step, the most expensive part is the generation of the search direction v_n^{MO}. In view of (6.2.3), (6.2.10), and (6.4.4), we have

$$\psi_n^{\text{MR}}(A)r_0 = r_n^{\text{MCR}} = r_{n-1}^{\text{MCR}} - \rho_{n-1}^{\text{MCR}} A v_n^{\text{MCR}}$$
$$= r_n^{\text{CR}} = r_{n-1}^{\text{CR}} - \eta_n^{\text{CR}} A w_n^{\text{CR}}.$$

So, one may compute the search direction in question via

$$v_n^{\text{MCR}} = \frac{\eta_n^{\text{CR}}}{\rho_{n-1}^{\text{MCR}}} w_n^{\text{CR}},$$

as long as ρ_{n-1}^{MCR} does not vanish. We know from Lemma 2.5.4(c) that $\rho_{n-1}^{\text{MCR}} = 0$ just happens for the degenerate case $\psi_n(0) = 0$, a situation where we do not want to use the CR implementation anyway.

Now, Chandra's idea is roughly as follows: whenever ρ_{n-1}^{MCR} is "large enough" compute the search direction via the CR scheme, otherwise compute the search direction via the MCR scheme. A careful implementation of this approach saves about 4N flops at each step where $|\rho_{n-1}^{\text{MCR}}| > \varepsilon$. Here, ε denotes some prescribed tolerance.

Freund pointed out that basically the same idea works for the STOD scheme. Again, the degenerate case $\psi_n'(0) = 0$ is characterized by a vanishing stepsize $\rho_{n-1}^{\text{STOD}} = 0$ (see Cor. 2.6.4). This time, one simply substitutes v_n^{MO} by $\rho_{n-1}^{\text{STOD}} v_n^{\text{MO}}$ and Av_n^{MO} by $\rho_{n-1}^{\text{STOD}} Av_n^{\text{MO}}$ if $|\rho_{n-1}^{\text{STOD}}|$ is sufficiently large. In each such step, one saves about N flops.

Both modification techniques leave the "user" with the trouble of choosing an appropriate switching value ε. The problem is that a too small ε may cause numerical instabilities, whereas a too large ε does not save any flops.

6.5 Implementations Based on the Lanczos Basis

In this section we will discuss another two implementations of polynomial iteration methods. Again, the underlying inner product is given by $\langle \cdot, \cdot \rangle_{\text{GAL}}$. This time, we make use of Theorem 2.6.7 to generate the iterates for

(6.5.1) $$r_n^{\text{ME}} = \psi_n^{\text{ME}}(A)r_0,$$

and of Theorem 2.5.7 for the method

(6.5.2) $$r_n^{\text{MR}} = \psi_n^{\text{MR}}(A)r_0.$$

The resulting implementations SYMMLQ (for (6.5.1)) and MINRES (for (6.5.2)) are (in principle) due to Paige and Saunders [95]. They were the first to use the Lanczos algorithm to advantage. Beside the fact that the Lanczos process constitutes a very efficient way for computing a basis of the Krylov space (cf. Section 4.1)

$$\text{span}\{v_1, v_2, \ldots, v_n\} = \mathcal{K}_n(A; r_0), \quad v_j = \psi_{j-1}(A)r_0,$$

the tridiagonal matrix $\hat{J}_n$ (cf. 2.1.25) enters into the picture. The next crucial step of Paige and Saunders towards an effective and stable scheme was the utilization of the LQ factorization of $\hat{J}_n$. We will instead make use the QR factorization of $\hat{J}_n$ (cf. Section 2.3). This technique was quite successfully used by Saad and Schultz [108] in designing their GMRES algorithm. Later on it was used by Freund [49] in conjunction with complex symmetric systems. In view of this consideration we should name the implementation to be discussed by SYMMQR. We will, however, stick to the (famous) name SYMMLQ, because, after all both implementations compute the same quantities.

We stress that both SYMMLQ and MINRES compute "on the fly" the (existing) iterates connected to the method

(6.5.3) $$r_n^{\text{OR}} = \psi_n^{\text{OR}}(A)r_0.$$

This is a direct consequence of Theorem 2.7.1 and Theorem 2.7.2. In addition, SYMMLQ is, in view of Theorem 2.7.3, capable of generating the MR iterates.

The SYMMLQ Approach

In this section we devise a stable method for the computation of the iterates implicitly defined by (6.5.1), (6.5.2), and (6.5.3). The approach is based on the representation of the scaled Hermite kernel polynomial ψ_n^{ME} in Theorem 2.6.7, on the representation of the scaled orthogonal residual polynomial ψ_n^{OR} in Theorem 2.7.1, and on the representation of the scaled kernel polynomials in Theorem 2.7.3. We have for the minimal error part

$$\begin{aligned} r_n^{\text{ME}} &= \psi_n^{\text{ME}}(A)r_0 \\ &= \psi_{n-1}^{\text{ME}}(A)r_0 - \hat{\eta}_n^{\text{ME}} A w_n^{\text{ME}}(A)r_0 \\ &= r_{n-1}^{\text{ME}} - \hat{\eta}_n^{\text{ME}} A w_n^{\text{ME}}, \end{aligned}$$

where, for convenience,

6.5 Implementations Based on the Lanczos Basis

$$w_n^{\text{ME}} := w_n^{\text{ME}}(A) r_0$$
$$= c_n w_n(A) r_0 + s_n \psi_n(A) r_0$$
$$= c_n w_n + s_n v_{n+1},$$

and

$$w_n := w_n(A) r_0$$
$$= -s_{n-1} w_{n-1}(A) r_0 + c_{n-1} \psi_{n-1}(A) r_0$$
$$= -s_{n-1} w_{n-1} + c_{n-1} v_n.$$

Moreover, the corresponding iterate looks like

$$x_n^{\text{ME}} = x_{n-1}^{\text{ME}} + \hat{\eta}_n^{\text{ME}} w_n^{\text{ME}}.$$

The Galerkin part may be computed, for $c_n \neq 0$, according to

$$r_n^{\text{OR}} = \psi_n^{\text{OR}}(A) r_0$$
$$= \psi_{n-1}^{\text{ME}}(A) r_0 - \frac{\hat{\eta}_n^{\text{ME}}}{c_n} A w_n(A) r_0$$
$$= r_{n-1}^{\text{ME}} - \frac{\hat{\eta}_n^{\text{ME}}}{c_n} A w_n,$$

and

$$x_n^{\text{OR}} = x_{n-1}^{\text{ME}} + \frac{\hat{\eta}_n^{\text{ME}}}{c_n} w_n.$$

Actually, in view of Theorem 2.7.3, one may in addition compute the minimal residual iterates (6.5.2) via

$$r_n^{\text{MR}} = \psi_n^{\text{MR}}(A) r_0$$
$$= s_n^2 \psi_{n-1}^{\text{MR}}(A) r_0 + c_n^2 \psi_{n-1}^{\text{ME}}(A) r_0 - c_n \eta_n^{\text{ME}} A w_n(A) r_0$$
$$= s_n^2 r_{n-1}^{\text{MR}} + c_n^2 r_{n-1}^{\text{ME}} - c_n \eta_n^{\text{ME}} A w_n,$$

which together with the fact $s_n^2 + c_n^2 = 1$ (cf. (2.3.4)) yields

$$x_n^{\text{MR}} = s_n^2 x_{n-1}^{\text{MR}} + c_n^2 x_{n-1}^{\text{ME}} + c_n \eta_n^{\text{ME}} w_n.$$

Notice that these computations are at "almost no extra cost". This means, no matrix-vector product and no inner product are required for the update of x_n^{MR}, and the vectors x_{n-1}^{ME} and w_n have to be computed anyway.

Finally, recall that the update formulae for the norm of the residuals and for $\hat{\eta}_j^{\text{ME}}$ are stated in Cor. 2.6.3, Theorem 2.7.1(b), and Cor. 2.5.3, respectively.

Theorem 6.5.1 *Let v_n denote the Lanczos vectors generated by the Lanczos method starting with $\beta_1 v_1 = r_0$. Furthermore, let the QR factorization $\{Q_n, R_n(r_{i,j}), G_n(c_n, s_n); \hat{J}_n\}$ (cf. (2.3.6)) be given.*

(a) The nth ME iterate is given by

$$x_n^{\text{ME}} = x_{n-1}^{\text{ME}} + \hat{\eta}_n^{\text{ME}}(c_n w_n + s_n v_{n+1}), \quad x_0^{\text{ME}} = x_0,$$

where

$$\hat{\eta}_n^{\text{ME}} = -(r_{3n}\hat{\eta}_{n-2}^{\text{ME}} + r_{2n}\hat{\eta}_{n-1}^{\text{ME}})/r_{1n}, \quad \hat{\eta}_0^{\text{ME}} = 0, \quad \hat{\eta}_1^{\text{ME}} = \beta_1/r_{11},$$

and

$$w_n = -s_{n-1}w_{n-1} + c_{n-1}v_n, \quad w_1 = v_1.$$

(b) If $c_n \neq 0$, then the nth OR iterate is given by

$$x_n^{\text{OR}} = x_{n-1}^{\text{ME}} + \hat{\eta}_n^{\text{ME}} w_n/c_n.$$

If $c_n = 0$, then x_n^{OR} does not exist.

(c) The nth MR iterate is given by

$$x_n^{\text{MR}} = s_n^2 x_{n-1}^{\text{MR}} + c_n^2 x_{n-1}^{\text{ME}} + c_n \eta_n^{\text{ME}} w_n, \quad x_0^{\text{MR}} = x_0.$$

(d) The norm of the corresponding residuals is given by

$$\|r_{n-2}^{\text{ME}}\|_2 = \sqrt{(r_{1,n-1}\hat{\eta}_{n-1}^{\text{ME}})^2 + (r_{3,n}\hat{\eta}_{n-2}^{\text{ME}})^2},$$

$$\|r_n^{\text{OR}}\|_2 = \beta_{n+1}\left|s_{n-1}\hat{\eta}_{n-1}^{\text{ME}} + c_{n-1}\hat{\eta}_n^{\text{ME}}/c_n\right|,$$

$$\|r_n^{\text{MR}}\|_2 = \beta_1 \prod_{j=1}^{n} |s_j|.$$

6.5 Implementations Based on the Lanczos Basis

MATLAB Implementation of SYMMLQ

All residual norms are calculated by update formulae. As a safeguard one should compare these quantities, at the end of the iteration, with the "true residuals" $\|f - Ax_n^{\mathrm{OR}}\|_2$, $\|f - Ax_n^{\mathrm{ME}}\|_2$, and $\|f - Ax_n^{\mathrm{MR}}\|_2$, respectively. Observe that the update formula for the norm of the nth ME residual involves quantities which are produced at step $n+2$ of the iteration.

Next, an operation count for SYMMLQ is presented. To have a fair comparison to other methods, we first ignore the flops associated with the computation of xMR.

Work per iteration and storage requirements of SYMMLQ:

- Matrix-vector products: `matvec(A,v)`   1
- dot products: `v'*Av, norm(v_hat)`   2
- additional flops: `Av-alpha*v-beta*v_old,`
  `v_hat/beta, -s_old*w+c_old*v,`
  `xME+(eta*c)*w+(eta*s)*v`   12N
- vectors to be stored: `v, v_hat, v_old, Av,`
  `w, xME`   6

In the next table we indicate the additional work caused by the computation of the MR iterates. To this end, observe that the update formulae for both xME and xMR involve the quantity `c*eta*w`.

Additional work and storage for SYMMLQ with MR phase:

- Matrix-vector products: 0
- dot products: 0
- additional flops: `s^2*xMR+c^2*xME+c*eta*w`   4N
- vectors to be stored: `xMR`   1

```
function [xOR,norm_rOR,xME,norm_rME,xMR,norm_rMR]=...
  SYMMLQ(A,f,x0,n_max,tol)
%
% computes the solution of Ax=f using the SYMMLQ scheme.
%
% startvector:                          x0
% maximal number of iterations:         n_max
% desired accuracy of the residual:     tol
%
% uses: matvec
%

%%initialize
  n=1; N=length(x0);
  v_hat=f-matvec(A,x0); beta=norm(v_hat);
  v=v_hat/beta; v_old=zeros(N,1);
  c=1; c_old=1; s_old=0; s=0; r1=1;
  xME=x0; xMR=x0; w=zeros(N,1);
  norm_rMR=beta; norm_r0=beta;  eta=-1; eta_old=0;

  while (n < n_max+1) & (norm_rMR/norm_r0 > tol)
     n=n+1;
%%Lanczos
     Av=matvec(A,v); alpha=v'*Av;
     v_hat=Av-alpha*v-beta*v_old;
     beta_old=beta; beta=norm(v_hat); v_old=v;
%%QR factorization
     c_oold=c_old; c_old=c; s_oold=s_old; s_old=s; r1_old=r1;
     r1_hat=c_old*alpha-c_oold*s_old*beta_old;
     r1    =sqrt(r1_hat^2+beta^2);
     r2    =s_old*alpha+c_oold*c_old*beta_old;
     r3    =s_oold*beta_old;
%%Givens rotation
     c=r1_hat/r1;
     s=beta/r1;
%%update
     w=-s_old*w+c_old*v; v=v_hat/beta;
     eta_oold=eta_old; eta_old=eta;
     eta=-(r3*eta_oold+r2*eta_old)/r1;

     xMR     =s^2*xMR+c^2*xME+c*eta*w;
     norm_rMR=norm_rMR*abs(s);
     xME     =xME+eta*(c*w+s*v); %wME  =c*w+s*v;
     norm_rME=sqrt((r1_old*eta_old)^2+(r3*eta_oold)^2);
  end %while

  if abs(c) > eps
     xOR     =xME+eta*w/c;
     norm_rOR=beta*abs(s_old*eta_old+c_old*eta/c);
  end;

return;
```

6.5 Implementations Based on the Lanczos Basis

Since, at least in theory, the norm of the MR residual is smaller than the other, we use this quantity as the stopping criteria. Furthermore, note that the vector wME need not to be stored. wME is only needed for the computation of xME. Also, it is easy to see that one could get rid of either v_hat or v_old if there would be a swap subroutine (cf. the comment for the STOD implementation).

The MINRES Approach

In this section we derive the MINRES implementation of (6.5.2) and (6.5.3). This time we make use of Theorem 2.5.7(b) for the computation of the scaled kernel polynomials ψ_n^{MR} and of Theorem 2.7.2(a) for the generation of the orthogonal residual polynomials ψ_n^{OR}. Starting with the minimal residual part, we obtain

$$\begin{aligned} r_n^{MR} &= \psi_n^{MR}(A)r_0 \\ &= \psi_{n-1}^{MR}(A)r_0 - c_n \hat{\eta}_{n-1}^{MR} A w_n^{MR}(A)r_0 \\ &= r_{n-1}^{MR} - c_n \hat{\eta}_{n-1}^{MR} A w_n^{MR}, \end{aligned}$$

where

$$\begin{aligned} w_n^{MR} &:= w_n^{MR}(A)r_0 \\ &= \left(\psi_{n-1}(A)r_0 - r_{3n} w_{n-2}^{MR}(A)r_0 - r_{2n} w_{n-1}^{MR}(A)r_0\right)/r_{1n} \\ &= \left(v_n - r_{3n} w_{n-2}^{MR} - r_{2n} w_{n-1}^{MR}\right)/r_{1n}. \end{aligned}$$

The corresponding iterates are computed by

$$x_n^{MR} = x_{n-1}^{MR} + c_n \hat{\eta}_{n-1}^{MR} w_n^{MR}.$$

The Galerkin case reduces to

$$\begin{aligned} r_n^{OR} &= \psi_n^{OR}(A)r_0 \\ &= \psi_n^{MR}(A)r_0 - \frac{s_n^2}{c_n} \hat{\eta}_{n-1}^{MR} A w_n^{MR}(A)r_0 \\ &= r_n^{MR} - \frac{s_n^2}{c_n} \hat{\eta}_{n-1}^{MR} A w_n^{MR} \end{aligned}$$

and, for $c_n \neq 0$,
$$x_n^{\text{OR}} = x_n^{\text{MR}} + \frac{s_n^2}{c_n}\hat{\eta}_{n-1}^{\text{MR}}w_n^{\text{MR}}.$$

Finally, recall that the update formulae for the norm of the residuals and for $\hat{\eta}_j^{\text{MR}}$ are stated in Cor. 2.5.3 and Theorem 2.7.2(c), respectively.

Theorem 6.5.2 *Let v_n denote the Lanczos vectors generated by the Lanczos method starting with $\beta_1 v_1 = r_0$. Furthermore, let the QR factorization $\{Q_n, R_n(r_{i,j}), G_n(c_n, s_n); \mathring{J}_n\}$ (cf. (2.3.6)) be given.*

(a) The nth MR iterate is given by
$$x_n^{\text{MR}} = x_{n-1}^{\text{MR}} + c_n\hat{\eta}_{n-1}^{\text{MR}}w_n^{\text{MR}}, \quad x_0^{\text{MR}} = x_0,$$

where
$$\hat{\eta}_{n-1}^{\text{MR}} = -s_{n-1}\hat{\eta}_{n-2}^{\text{MR}}, \quad \hat{\eta}_1^{\text{MR}} = -\beta_1 s_1,$$

and
$$w_n^{\text{MR}} = \left(v_n - r_{3n}w_{n-2}^{\text{MR}} - r_{2n}w_{n-1}^{\text{MR}}\right)/r_{1n},$$
$$w_1^{\text{MR}} = v_1/r_{11}, \quad w_2^{\text{MR}} = \left(v_2 - r_{22}w_1^{\text{MR}}\right)/r_{12}.$$

(b) If $c_n \neq 0$, then the nth OR iterate is given by
$$x_n^{\text{OR}} = x_n^{\text{MR}} + s_n^2\hat{\eta}_{n-1}^{\text{MR}}w_n^{\text{MR}}/c_n.$$

If $c_n = 0$, then x_n^{OR} does not exist.

(c) The norm of the corresponding residuals is given by
$$\|r_n^{\text{MR}}\|_2 = \beta_1 \prod_{j=1}^n |s_j|, \quad \|r_n^{\text{OR}}\|_2 = \|r_n^{\text{MR}}\|_2/|c_n|.$$

6.5 Implementations Based on the Lanczos Basis

MATLAB Implementation of MINRES

```
function [xMR,norm_rMR,xOR,norm_rOR]=MINRES(A,f,x0,n_max,tol)
%
% computes the solution of Ax=f using the MINRES scheme.
%
% startvector:                        x0
% maximal number of iterations:       n_max
% desired accuracy of the residual: tol
%
% uses: matvec
%

%%initialize
   n=1; N=length(x0);
   v=zeros(N,1); v_hat=f-matvec(A,x0); beta=norm(v_hat);
   c=1; c_old=1; s_old=0; s=0;
   w=zeros(N,1); w_old=w; eta=beta;
   xMR=x0; norm_rMR=beta; norm_r0=beta;

   while (n < n_max+1) & (norm_rMR/norm_r0 > tol)
      n=n+1;

%%Lanczos
      v_old=v;
      v=v_hat/beta; Av=matvec(A,v); alpha=v'*Av;
      v_hat=Av-alpha*v-beta*v_old;
      beta_old=beta; beta=norm(v_hat);

%%QR factorization
      c_oold=c_old; c_old=c; s_oold=s_old; s_old=s;

      r1_hat=c_old*alpha-c_oold*s_old*beta_old;
      r1    =sqrt(r1_hat^2+beta^2);
      r2    =s_old*alpha+c_oold*c_old*beta_old;
      r3    =s_oold*beta_old;

%%Givens rotation
      c=r1_hat/r1;
      s=beta/r1;

%%update
      w_oold=w_old; w_old=w;
      w=(v-r3*w_oold-r2*w_old)/r1;
      xMR=xMR+c*eta*w; norm_rMR=norm_rMR*abs(s);
      eta=-s*eta;

   end %while

   if abs(c) > eps
      xOR      =xMR-s*eta*w/c;
      norm_rOR=norm_rMR/abs(c);
   end;

return;
```

The operation count for MINRES is summarized in the next table.

Work per iteration and storage requirements of `MINRES`:

- Matrix-vector products: `matvec(A,v)`   1
- dot products: `v'*Av, norm(v_hat)`   2
- additional flops: `Av-alpha*v-beta*v_old,`
  `(v-r3*w_ool d-r2*w_old)/r1,`
  `xMR+c*eta*w, v_hat/beta`   12N
- vectors to be stored: `v, v_hat, v_old, Av,`
  `w, w_old, xMR`   7

It is possible to save some computational costs by working with the vector `r1*w` instead of `w`. Furthermore, notice that `v_old` and `w_ool d` are only "temporary" vectors, that is, they can share one storage place. In fact, one could save one more storage place if a `swap` facility would be available.

6.6 Residual Smoothing

Stopping criteria for iterative methods involve (in general) the Euclidian norm of the residuals r_n. Thus it is desirable that $\|r_n\|_2$ converges "smoothly" to zero. As is apparent from Figure 6.3.7 in Example 6.3.6 the CG (OR) scheme may exhibit very irregular residual norm behavior when applied to indefinite systems.

An elegant approach to "repair this shortcoming" has been proposed by Schönauer [111] and Weiss [131]. Their technique has been generalized in a recent paper by Zhou and Walker [138].

Let us briefly outline the approach devised in [111] and [131]. To this end let $\{x_n\}$ denote a sequence of iterates with "nervous" residual norm behavior. The idea is to generate an auxiliary sequence $\{\hat{x}_n\}$ from $\{x_n\}$ such that the new residuals $\hat{r}_n := f - A\hat{x}_n$ have monotone decreasing Euclidian norms. This goal may be achieved by the following "repair procedure"

(6.6.1) $$\hat{x}_0 = x_0$$
$$\hat{x}_j = (1 - \eta_j)\hat{x}_{j-1} + \eta_j x_j, \quad j \geq 1,$$

6.7 A "Non-Feasible" Approach

where η_j is chosen to minimize $\|\hat{r}_j\|_2 = \|\hat{r}_{j-1} + \eta_j(r_j - \hat{r}_{j-1})\|_2$. Clearly, the unique solution of this one dimensional approximation problem is given by

$$(6.6.2) \qquad \eta_j = -\frac{\hat{r}_{j-1}^T(r_j - \hat{r}_{j-1})}{\|r_j - \hat{r}_{j-1}\|_2^2}.$$

In accordance with Zhou and Walker, we refer to this technique as *minimal residual smoothing*.

Notice that $\hat{x}_n$ is a convex combination of $\hat{x}_{n-1}$ and x_n. Since the iterates produced by the scaled kernel polynomials ψ_j^{MR} and iterates associated with the orthogonal residual polynomials ψ_j^{OR} share this property (cf. Theorem 2.7.2(b))

$$x_j^{\text{MR}} = (1 - c_j^2)x_{j-1}^{\text{MR}} + c_j^2 x_j^{\text{OR}},$$

and since x_j^{MR} minimizes the residual with respect to the "full" Krylov subspace $\mathcal{K}_n$ we conclude

---

Theorem 6.6.1 *The polynomial iteration method based on scaled kernel polynomials (2.5.5) may be obtained from the polynomial iteration method based on orthogonal residual polynomials (2.4.3) by the minimal residual smoothing technique defined in (6.6.1) and (6.6.2).*

6.7 A "Non-Feasible" Approach

Let us briefly discuss why it is not feasible to look for iterates x_n^{MtE} defined by the minimal Euclidian error property over the space $x_0 + \mathcal{K}_n(A, r_0)$

$$(6.7.1) \qquad \|x_* - x_n^{\text{MtE}}\|_2 = \min\{\|x_* - x\|_2 : x \in x_0 + \mathcal{K}_n(A, r_0)\}.$$

Recall that one may compute a basis for $\mathcal{K}_n(A, r_0)$ by the Lanczos process (see Section 4.1)

$$\text{span}\{v_1, v_2, \ldots, v_n\} = \mathcal{K}_n(A, r_0).$$

Let us collect together the vectors v_j in (cf. (4.1.3))

$$V_n := \begin{bmatrix} v_1 & v_2 & \cdots & v_n \end{bmatrix}.$$

Thus, any element x from the shifted Krylov subspace $x_0 + \mathcal{K}_n(A, r_0)$ may be represented as
$$x = x_0 + V_n y, \quad y \in \mathbb{R}^n.$$
To compute x_n^{MtE}, we have

(6.7.2)
$$\begin{aligned}\|x_* - x\|_2^2 &= \|\varepsilon_0 - V_n y\|_2^2 \\ &= \|\varepsilon_0\|_2^2 - 2y^{\mathrm{T}} V_n^{\mathrm{T}} \varepsilon_0 + y^{\mathrm{T}} y.\end{aligned}$$

Hence, $x_n^{\mathrm{MtE}} = x_0 + V_n y_n^{\mathrm{MtE}}$ solves (6.7.1) for
$$y_n^{\mathrm{MtE}} = V_n^{\mathrm{T}} \varepsilon_0.$$

Obviously, the computation of the stationary point of (6.7.2) requires the knowledge of the solution x_*, which is not feasible (compare also Stoer [118, p. 548] and Deuflhard [31, §2.1]). It should be mentioned that there is, however, one exception to this rule. For Craig's version of the normal equation, the minimal true error approach is very well feasible (see Section 6.8).

In order to judge the various iterative methods, we will compare their respective errors $\|\varepsilon_n\|_2$ with the best obtainable error $\|\varepsilon_n^{\mathrm{MtE}}\|_2$. To this end, note that

$$\begin{aligned}\|\varepsilon_n^{\mathrm{MtE}}\|_2 = \|x_* - x_n^{\mathrm{MtE}}\|_2 &= \sqrt{\|\varepsilon_0\|_2^2 - \|V_n^{\mathrm{T}} \varepsilon_0\|_2^2} \\ &= \sqrt{\|\varepsilon_{n-1}^{\mathrm{MtE}}\|_2^2 - (v_n^{\mathrm{T}} \varepsilon_0)^2}.\end{aligned}$$

An implementation of this formula may suffer from cancellation errors. In order to keep the issues of interest clear, we will compute $\|\varepsilon_n^{\mathrm{MtE}}\|_2$ by using MATHEMATICA with high precision calculations.

MATHEMATICA Computation of the Minimal Error

If one explicitly defines the input parameter A, f, xstar, x0, with prec decimal digits, e.g.,

```
prec=50;   x0=Table[Random[Real,{-1,1},prec], {N}];,
```

then the call

```
normMtE=MtE[A,f,xstar,x0,nmax,tol]
```

will compute normMtE in an arithmetic with prec decimal digits.

6.8 Implementations Based on Normal Equations

```
MtE[A_,f_,xstar_,x0_,nmax_,tol_] :=

(*computes the Euclidian norm normMtE of the    *)
(*minmal error with respect to the Krylov space *)
(*generated by A, f, and x0.                    *)

(*maximal number of iterations:   nmax          *)
(*desired accuracy of the error:  tol           *)

(*uses: Matvec, Norm                            *)

Block[{r0,e0,v,vold,vhat,alpha,beta,norme,k,nn},

(*initialize*)
   nn=Length[x0];
   r0=f-Matvec[A,x0]; e0=xstar-x0;
   vold=Table[0,{nn}]; vhat=r0;
   beta=Norm[vhat];
   norme=Norm[e0]; norme0=norme;
   normMtE={}; AppendTo[normMtE,norme];

   For[k=2,((k<=nmax+1) && (norme/norme0>tol)),k++,

    (*Lanczos*)
       v=vhat/beta; Av=Matvec[A,v]; alpha=v . Av;
       vhat=Av - alpha v - beta vold;
       beta=Norm[vhat]; vold=v;

    (*norm*)
       norme=Sqrt[norme^2 - (v . e0)^2];
       AppendTo[normMtE,norme];

   ]; (*end For*)
(*output*)
   Return[normMtE];
];
```

6.8 Implementations Based on Normal Equations

In this section we will investigate methods for solving the normal equations, which come in two versions
$$A^2 x = Af$$
and
(6.8.1) $$A^2 y = f, \quad \text{with} \quad Ay = x.$$
It will turn out that in both cases we have to look for iterates in the shifted Krylov space $x_0 + A\mathcal{K}_n(A^2, r_0)$. This time the basis for $A\mathcal{K}_n(A^2, r_0)$ is generated by a bidiagonalization procedure instead of a tridiagonalization procedure as discussed in Section 4.1.

The Golub/Kahan Bidiagonalization

In principle one could compute the basis

(6.8.2) $$\text{span}\{v_1^{\text{NE}}, v_2^{\text{NE}}, \ldots, v_n^{\text{NE}}\} = \mathcal{K}_n(A^2, Ar_0)$$

by applying the Lanczos process (cf. Section 4.1) onto A^2 starting with Ar_0. However, we would like to discuss an alternative method. This method was suggested by Golub and Kahan [68]. The trick is to apply the Lanczos process to the matrix

$$\tilde{A} := \begin{bmatrix} I_N & A \\ A & 0 \end{bmatrix} \quad \text{starting with} \quad \tilde{r}_0 := \begin{bmatrix} r_0 \\ 0 \end{bmatrix}.$$

It is straightforward to verify that

$$\mathcal{K}_{2n}(\tilde{A}, \tilde{r}_0) = \begin{bmatrix} \mathcal{K}_n(A^2, r_0) \\ \mathcal{K}_n(A^2, Ar_0) \end{bmatrix}.$$

Hence, applying $2n$ steps of the Lanczos process to $\tilde{A}$ will generate the desired basis for $\mathcal{K}_n(A^2, Ar_0)$. Following Algorithm 4.1.1, the main ingredients of the procedure look like

$$\alpha_j = \begin{bmatrix} u_j \\ v_j \end{bmatrix} \tilde{A} \begin{bmatrix} u_j^T & v_j^T \end{bmatrix},$$

(6.8.3) $$\begin{bmatrix} \hat{u}_{j+1} \\ \hat{v}_{j+1} \end{bmatrix} = (\tilde{A} - \alpha_j I_{2N}) \begin{bmatrix} u_j \\ v_j \end{bmatrix} - \beta_j \begin{bmatrix} u_{j-1} \\ v_{j-1} \end{bmatrix},$$

$$\beta_{j+1} = \left\| \begin{bmatrix} \hat{u}_{j+1} \\ \hat{v}_{j+1} \end{bmatrix} \right\|_2, \quad \begin{bmatrix} u_{j+1} \\ v_{j+1} \end{bmatrix} = \frac{1}{\beta_{j+1}} \begin{bmatrix} \hat{u}_{j+1} \\ \hat{v}_{j+1} \end{bmatrix}.$$

Due to the special structure of $\tilde{A}$ and $\tilde{r}_0$ the procedure simplifies. A straightforward calculation shows that

$$\alpha_{2j} = 0, \quad \hat{u}_{2j} = 0, \quad \hat{v}_{2j} = Au_{2j-1} - \beta_{2j-1} v_{2j-2}, \quad \beta_{2j} = \|\hat{v}_{2j}\|_2,$$

and

$$\alpha_{2j+1} = 1, \quad \hat{u}_{2j+1} = Av_{2j} - \beta_{2j} u_{2j-1}, \quad \hat{v}_{2j+1} = 0, \quad \beta_{2j+1} = \|\hat{u}_{2j+1}\|_2.$$

6.8 Implementations Based on Normal Equations

With the setting
$$\hat{u}_j^{\text{NE}} := \hat{u}_{2j-1}, \quad u_j^{\text{NE}} := u_{2j-1},$$
$$\hat{v}_j^{\text{NE}} := \hat{v}_{2j}, \quad v_j^{\text{NE}} := u_{2j},$$

and
$$\alpha_j^{\text{NE}} := \beta_{2j} = \|\hat{v}_j^{\text{NE}}\|_2, \quad \beta_j^{\text{NE}} := \beta_{2j-1} = \|\hat{u}_j^{\text{NE}}\|_2,$$

we may reformulate the key equation for the Lanczos process as
$$\hat{u}_{j+1}^{\text{NE}} = A v_j^{\text{NE}} - \alpha_j^{\text{NE}} u_j^{\text{NE}}, \quad \hat{u}_1^{\text{NE}} = r_0,$$
$$\hat{v}_{j+1}^{\text{NE}} = A u_{j+1}^{\text{NE}} - \beta_{j+1}^{\text{NE}} v_j^{\text{NE}}, \quad \hat{v}_1^{\text{NE}} = A u_1^{\text{NE}}.$$

or, equivalently, as

(6.8.4)
$$A v_j^{\text{NE}} = \beta_{j+1}^{\text{NE}} u_{j+1}^{\text{NE}} + \alpha_j^{\text{NE}} u_j^{\text{NE}}, \quad r_0 = \beta_1^{\text{NE}} u_1^{\text{NE}},$$
$$A u_{j+1}^{\text{NE}} = \alpha_{j+1}^{\text{NE}} v_{j+1}^{\text{NE}} + \beta_{j+1}^{\text{NE}} v_j^{\text{NE}}, \quad A u_1^{\text{NE}} = \alpha_1^{\text{NE}} \hat{v}_1^{\text{NE}}.$$

This time the bidiagonal matrices

(6.8.5)
$$J_n^{\text{NE}} = \begin{bmatrix} \alpha_1^{\text{NE}} & 0 & 0 & \cdots & 0 \\ \beta_2^{\text{NE}} & \alpha_2^{\text{NE}} & \ddots & \ddots & \vdots \\ 0 & \ddots & \ddots & \ddots & 0 \\ \vdots & \ddots & \ddots & \ddots & 0 \\ 0 & \cdots & 0 & \beta_n^{\text{NE}} & \alpha_n^{\text{NE}} \end{bmatrix}$$

and

(6.8.6)
$$\hat{J}_n^{\text{NE}} = \begin{bmatrix} J_n^{\text{NE}} \\ \beta_{n+1}^{\text{NE}} e_n^{\text{T}} \end{bmatrix}$$

enter into the picture. Let us collect the vectors u_j^{NE}, v_j^{NE} in

(6.8.7)
$$U_n^{\text{NE}} = [u_1^{\text{NE}} \quad u_2^{\text{NE}} \quad \cdots \quad u_n^{\text{NE}}],$$
$$V_n^{\text{NE}} = [v_1^{\text{NE}} \quad v_2^{\text{NE}} \quad \cdots \quad v_n^{\text{NE}}],$$

respectively. In accordance with (4.1.4) and (4.1.5), the equations (6.8.4) read in compact matrix notation

(6.8.8)
$$A V_n^{\text{NE}} = U_n^{\text{NE}} J_n^{\text{NE}} + \beta_{n+1}^{\text{NE}} u_{n+1}^{\text{NE}} e_n^{\text{T}} = U_{n+1}^{\text{NE}} \hat{J}_n^{\text{NE}}, \quad r_0 = \beta_1^{\text{NE}} U_{n+1}^{\text{NE}} e_1,$$
$$A U_{n+1}^{\text{NE}} = V_n^{\text{NE}} (\hat{J}_n^{\text{NE}})^{\text{T}} + \alpha_{n+1}^{\text{NE}} v_{n+1}^{\text{NE}} e_{n+1}^{\text{T}}.$$

Also, we have by construction

(6.8.9)
$$(V_n^{\text{NE}})^{\text{T}} V_n^{\text{NE}} = I_n, \quad \text{and} \quad (U_n^{\text{NE}})^{\text{T}} U_n^{\text{NE}} = I_n$$

which together with (6.8.8) yields

$$(U_n^{\text{NE}})^T A V_n^{\text{NE}} = J_n^{\text{NE}}.$$

In other words, the Lanczos process (6.8.3) results in a *bidiagonalization method*.

QR Factorization; Bidiagonal Case

Not surprisingly, in developing stable methods for the iterative solution of the normal equations, we will again make use of a QR factorization. This time we are looking for the upper triangular factor

$$(6.8.10) \qquad R_n^{\text{NE}} = \begin{bmatrix} r_{11}^{\text{NE}} & r_{22}^{\text{NE}} & 0 & 0 & \cdots & 0 \\ 0 & r_{12}^{\text{NE}} & r_{23}^{\text{NE}} & \ddots & \ddots & \vdots \\ \vdots & \ddots & r_{13}^{\text{NE}} & \ddots & \ddots & 0 \\ \vdots & & \ddots & \ddots & \ddots & 0 \\ \vdots & & & \ddots & \ddots & r_{2n}^{\text{NE}} \\ 0 & \cdots & \cdots & \cdots & 0 & r_{1n}^{\text{NE}} \end{bmatrix}$$

in the QR factorization

$$(6.8.11) \qquad Q_n^{\text{NE}} \hat{J}_n^{\text{NE}} = \begin{bmatrix} R_n^{\text{NE}} \\ 0 \end{bmatrix}$$

of the bidiagonal matrix $\hat{J}_n^{\text{NE}}$ defined in (6.8.6). Since $\hat{J}_n^{\text{NE}}$ has more zero entries than $\hat{J}_n$, the formulae in Lemma 2.3.1 simplify.

Lemma 6.8.1 *Let* $\{\hat{J}_n^{\text{NE}}, (\alpha_n^{\text{NE}}, \beta_n^{\text{NE}}); \langle \cdot, \cdot \rangle_{\text{GAL}}\}$ *be given. The upper triangular factor* R_n^{NE} *(cf. (6.8.5)) of the QR factorization (6.8.1) and the Givens rotation* G_n^{NE} *(cf. (2.3.4)) are given by*

$$r_{2n}^{\text{NE}} = s_{n-1}^{\text{NE}} \alpha_n^{\text{NE}}, \quad n \geq 2$$

$$r_{1n}^{\text{NE}} = \sqrt{(\hat{r}_{1n}^{\text{NE}})^2 + (\beta_{n+1}^{\text{NE}})^2}, \quad n \geq 1; \quad \hat{r}_{1n}^{\text{NE}} = \begin{cases} \alpha_1^{\text{NE}}, & n = 1, \\ c_{n-1}^{\text{NE}} \alpha_n^{\text{NE}}, & n \geq 2; \end{cases}$$

$$c_n^{\text{NE}} = \frac{\hat{r}_{1n}^{\text{NE}}}{r_{1n}^{\text{NE}}}, \quad s_n^{\text{NE}} = \frac{\beta_{n+1}^{\text{NE}}}{r_{1n}^{\text{NE}}}, \quad n \geq 1.$$

6.8 Implementations Based on Normal Equations

The LSQR Approach

Perhaps the most obvious indefinite iterative method is the application of the conjugate gradient iteration to the normal equations $A^2 x = Af$. Actually, this idea goes back to the original paper by Hestenes and Stiefel [81]. Of course A^2 is never formed explicitly. This method is known by the abbreviation CGNR. Here, we will outline the LSQR implementation of Paige and Saunders [96]. It is mathematically equivalent to CGNR, *i.e.*, in exact arithmetic it produces the same iterates, but has better numerical properties.

Before we actually derive the LSQR scheme, we will investigate some of its properties. To this end, let us equip the quantities associated with the normal equations with a "hat", e.g., $\hat{A} := A^2$ and $\hat{f} := Af$. Then applying the CG method to the normal equations $\hat{A}\hat{x} = \hat{f}$ leads to the polynomial iteration method

$$\hat{r}_n = \hat{f} - \hat{A}\hat{x}_n = \hat{\psi}_n^{\mathrm{OR}}(\hat{A})\hat{r}_0, \quad \hat{r}_0 = Ar_0,$$

where

$$\|\hat{\varepsilon}_n\|_{\hat{A}} = \|\hat{\psi}_n^{\mathrm{OR}}(\hat{A})\hat{\varepsilon}_0\|_{\hat{A}} = \min\left\{\|p(\hat{A})\hat{\varepsilon}_0\|_{\hat{A}} : p \in \Pi_n, \ p(0) = 1\right\}.$$

Equivalently, $\hat{\psi}_n^{\mathrm{OR}}$ is defined by the Galerkin condition (cf. 2.4.7)

$$0 = \hat{r}_0^{\mathrm{T}} \hat{\psi}_k^{\mathrm{OR}}(\hat{A}) \hat{\psi}_l^{\mathrm{OR}}(\hat{A}) \hat{r}_0 = r_0^{\mathrm{T}} \hat{\psi}_k^{\mathrm{OR}}(\hat{A}) \hat{A} \hat{\psi}_l^{\mathrm{OR}}(\hat{A}) r_0, \quad k \ne l.$$

Thus, in view of Theorem 2.5.1, the polynomial $\hat{\psi}_n^{\mathrm{OR}}$ may be viewed as the scaled kernel polynomial with respect to the inner product

$$\langle p, q \rangle = r_0^{\mathrm{T}} p(A^2) q(A^2) r_0$$

and consequently

(6.8.12)
$$\|f - A\hat{x}_n\|_2 = \|\hat{\psi}_n^{\mathrm{OR}}(A^2) r_0\|_2$$
$$= \min\left\{\|p(A^2) r_0\|_2 : p \in \Pi_n, \ p(0) = 1\right\}$$

with

$$\|x_* - \hat{x}_n\|_{\hat{A}} = \|f - A\hat{x}_n\|_2.$$

In other words, $x_n^{\mathrm{LSQR}} = \hat{x}_n$ minimizes as well the Euclidian norm of the residual (with respect to the original system $Ax = f$) over the affine Krylov space $x_0 + A\mathcal{K}_n(A^2, r_0)$. Note that the letter "R" in CGNR indicates the minimization of the residual.

The LSQR implementation exploits the minimization property (6.8.12) of the residual. Apart from the way it computes the basis vectors it is essentially a MINRES type implementation.

Let
$$x_n^{\text{LSQR}} = x_0 + V_n^{\text{NE}} y_n^{\text{LSQR}}$$
be represented in terms of the Lanczos matrix V_n^{NE} (cf. (6.8.7)). Then it follows from (6.8.8)
$$r_n^{\text{LSQR}} = f - A x_n^{\text{LSQR}} = r_0 - A V_n^{\text{NE}} y_n^{\text{LSQR}} = U_{n+1}^{\text{NE}} (\beta_1^{\text{NE}} e_1 - \hat{J}_n^{\text{NE}} y_n^{\text{LSQR}}).$$

Since the columns of U_{n+1}^{NE} are orthogonal (cf. 6.8.9) the determination of y_n^{LSQR} leads to the following least-squares problem

(6.8.13) $\quad \|\beta_1^{\text{NE}} e_1 - \hat{J}_n^{\text{NE}} y_n^{\text{LSQR}}\|_2 = \min\{\|\beta_1^{\text{NE}} e_1 - \hat{J}_n^{\text{NE}} y\|_2 : y \in R^n\}.$

To solve (6.8.13) we apply the QR factorization (6.8.11) of J_n^{NE} and obtain

$$\|\beta_1^{\text{NE}} e_1 - \hat{J}_n^{\text{NE}} y\|_2^2 = \|\beta_1^{\text{NE}} Q_n^{\text{NE}} e_1 - Q_n^{\text{NE}} \hat{J}_n^{\text{NE}} y\|_2^2$$
$$= \|\hat{y}_n^{\text{LSQR}} - R_n^{\text{NE}} y\|_2^2 + (\hat{\eta}_n^{\text{LSQR}})^2,$$

where
$$\begin{bmatrix} \hat{y}_n^{\text{LSQR}} \\ \hat{\eta}_n^{\text{LSQR}} \end{bmatrix} := \beta_1^{\text{NE}} Q_n^{\text{NE}} e_1.$$

Hence, the minimum of (6.8.13) is attained for the solution y_n^{LSQR} of

$$R_n^{\text{NE}} y = \hat{y}_n^{\text{LSQR}}.$$

This is a familiar problem. It appeared already, with MR instead of LSQR, in the derivation of the MINRES scheme (cf. Cor. 2.5.3). We have our next theorem.

6.8 Implementations Based on Normal Equations

Theorem 6.8.2 *Let v_n^{NE} (cf. (6.8.2)) denote the basis vectors generated by the bidiagonalization procedure (6.8.4) starting with $\beta_1^{NE} u_1^{NE} = r_0$. Furthermore, let the QR factorization $\{Q_n^{NE}, R_n^{NE}(r_{ij}^{NE}), G_n^{NE}(c_n^{NE}, s_n^{NE}); \hat{J}_n^{NE})\}$ (cf. Lemma 6.8.1) be given.*

(a) The nth LSQR iterate is given by

$$x_n^{LSQR} = x_{n-1}^{LSQR} + c_n^{NE} \hat{\eta}_{n-1}^{LSQR} w_n^{LSQR}, \quad x_0^{LSQR} = x_0,$$

where

$$\hat{\eta}_{n-1}^{LSQR} = -s_{n-1}^{NE} \hat{\eta}_{n-2}^{LSQR}, \quad \hat{\eta}_1^{LSQR} = -\beta_1^{NE} s_1^{NE},$$

$$w_n^{LSQR} = \left(v_n^{NE} - r_{2n}^{NE} w_{n-1}^{LSQR}\right)/r_{1n}^{NE}, \quad w_1^{LSQR} = v_1^{NE}/r_{11}^{NE}.$$

(b) The norm of the corresponding residual is given by

$$\|r_n^{LSQR}\|_2 = \beta_1^{NE} \prod_{j=1}^{n} |s_j^{NE}| = |s_n^{NE}| \, \|r_{n-1}^{LSQR}\|_2.$$

MATLAB Implementation of LSQR

The operation count for LSQR is listed in the next table.

Work per iteration and storage requirements of LSQR:

- Matrix-vector products: matvec(A,u), matvec(A,v) 2
- dot products: norm(u), norm(v) 2
- additional flops: Au-beta*v, v/alpha,
  Av-alpha*u, u/beta,
  (v-r2*w)/r1, xLSQR+c*eta*w 11N
- vectors to be stored: u, Au, v, w, xLSQR 5

```
function [xLSQR,norm_rLSQR]=LSQR(A,f,x0,n_max,tol)
%
% computes the solution of Ax=f using the LSQR scheme.
%
% startvector:                          x0
% maximal number of iterations:         n_max
% desired accuracy of the residual: tol
%
% uses: matvec
%

%%initialize
    n=1; N=length(x0);
    v=zeros(N,1);
    u=f-matvec(A,x0); beta=norm(u); u=u/beta;
    c=1; s=0;
    w=zeros(N,1); eta=beta;
    xLSQR=x0; norm_rLSQR=beta; norm_r0=beta;

    while (n < n_max+1) & (norm_rLSQR/norm_r0 > tol)
        n=n+1;

    %%Bidiagonalization
        Au=matvec(A,u);
        v=Au-beta*v; alpha=norm(v); v=v/alpha;
        Av=matvec(A,v);
        u=Av-alpha*u; beta=norm(u); u=u/beta;

    %%QR factorization
        r1_hat=c*alpha;
        r1     =sqrt(r1_hat^2+beta^2);
        r2     =s*alpha;

    %%Givens rotation
        c=r1_hat/r1;
        s=beta/r1;

    %%update
        w=(v-r2*w)/r1;
        xLSQR=xLSQR+c*eta*w;
        eta=-s*eta;

    %%compute norm
        norm_rLSQR=norm_rLSQR*abs(s);

    end %while

return;
```

Note that the LSQR implementation requires two matrix vector multiplications per step. Moreover, `Au` and `Av` can share one storage place.

6.8 Implementations Based on Normal Equations

The CRAIG Approach

Now, let us apply the standard conjugate gradient method to the second version of the normal equations $A^2 y = f$, $Ay = x$ (cf. (6.8.1)). The resulting algorithm CGNE is often referred to as *Craig's method* [24]. Again, we would like to discuss an implementation based on the bidiagonalization scheme (6.8.4). This method is due to Paige [94]. It is mathematically equivalent to CGNE, i.e., in exact arithmetic it would generate the same iterates.

If $\hat{x}_n$ denotes the nth iterate, then we have, with the setting $\hat{A} := A^2$, $Ay_* = x_*$, and $Ay_0 = x_0$,

$$\hat{r}_0 = f - \hat{A} y_0 = f - A x_0 = r_0 \quad \text{and} \quad \hat{\varepsilon}_n = y_* - \hat{x}_n.$$

This time the polynomial iteration method reads

$$\hat{r}_n = f - \hat{A} \hat{x}_n = \hat{\psi}_n^{\text{OR}}(\hat{A}) \hat{r}_0,$$

where

$$\|\hat{\varepsilon}_n\|_{\hat{A}} = \|\hat{\psi}_n^{\text{OR}}(\hat{A}) \hat{\varepsilon}_0\|_{\hat{A}} = \min \left\{ \|p(\hat{A}) \hat{\varepsilon}_0\|_{\hat{A}} : p \in \Pi_n,\ p(0) = 1 \right\}.$$

The residual polynomials $\hat{\psi}_n^{\text{OR}}$ satisfy the following orthogonality relation

$$0 = \hat{r}_0^{\text{T}} \hat{\psi}_k^{\text{OR}}(\hat{A}) \hat{\psi}_l^{\text{OR}}(\hat{A}) \hat{r}_0 = \varepsilon_0^{\text{T}} \hat{\psi}_k^{\text{OR}}(\hat{A}) \hat{A} \hat{\psi}_l^{\text{OR}}(\hat{A}) \varepsilon_0, \quad k \neq l.$$

So, in accordance with the CGNR approach, the polynomial $\hat{\psi}_n^{\text{OR}}$ may be viewed as the scaled kernel polynomial with respect to the inner product

$$\langle p, q \rangle = \varepsilon_0^{\text{T}} p(A^2) q(A^2) \varepsilon_0.$$

From Theorem 2.5.1 we deduce

(6.8.14)
$$\|x_* - \hat{x}_n\|_2 = \|\hat{\psi}_n^{\text{OR}}(A^2) \varepsilon_0\|_2$$
$$= \min \left\{ \|p(A^2) \varepsilon_0\|_2 : p \in \Pi_n,\ p(0) = 1 \right\}$$

with

$$\|y_* - \hat{x}_n\|_{\hat{A}} = \|x_* - \hat{x}_n\|_2.$$

In conclusion, CGNE differs from CGNR (cf. (6.8.12)) in that it minimizes the Euclidian norm of the error over the shifted Krylov subspace

$x_0 + A\mathcal{K}_n(A^2, r_0)$, instead of the Euclidian norm of the residual. Note that the letter "E" in CGNE indicates the minimization of the error.

So, we are in a "minimal true error situation" (cf. Section 6.7). This time, however, the iterates can be computed without knowing the solution. In accordance with (6.7.2) we deduce, for $x = x_0 + V_n^{\text{NE}} y$,

$$\text{(6.8.15)} \qquad \|x_* - x\|_2^2 = \|\varepsilon_0 - V_n^{\text{NE}} y\|_2^2 = \|\varepsilon_0\|_2^2 - 2y^{\text{T}}(V_n^{\text{NE}})^{\text{T}} \varepsilon_0 + y^{\text{T}} y.$$

Hence, $x_n^{\text{CRAIG}} = \hat{x}_n = x_0 + V_n^{\text{NE}} y_n^{\text{CRAIG}}$ minimizes (6.8.15) for

$$y_n^{\text{CRAIG}} = (V_n^{\text{NE}})^{\text{T}} \varepsilon_0.$$

In view of (6.8.8) and (6.8.9) the equation above for y_n^{CRAIG} can be modified as follows

$$\begin{aligned} J_n^{\text{NE}} y_n^{\text{CRAIG}} &= J_n^{\text{NE}} (V_n^{\text{NE}})^{\text{T}} \varepsilon_0 \\ &= (U_n^{\text{NE}})^{\text{T}} A \varepsilon_0 \\ &= \beta_1^{\text{NE}} e_1. \end{aligned}$$

Notice that the (unknown) initial error ε_0 has disappeared. The representation $y_n^{\text{CRAIG}} = \beta_1^{\text{NE}} (J_n^{\text{NE}})^{-1} e_1$ immediately yields an update formula for y_n^{CRAIG}. To see this, let $y_n^{\text{CRAIG}} =: (\xi_1, \xi_2, \ldots, \xi_n)^{\text{T}}$ and observe that by (6.8.5)

$$\xi_{j+1} = -\frac{\beta_{j+1}^{\text{NE}}}{\alpha_{j+1}^{\text{NE}}} \xi_j, \qquad \xi_1 = \frac{\beta_1^{\text{NE}}}{\alpha_1^{\text{NE}}}.$$

Moreover, the residuals are orthogonal

$$\begin{aligned} r_n^{\text{CRAIG}} &= f - A x_n^{\text{CRAIG}} \\ &= r_0 - A V_n^{\text{NE}} y_n^{\text{CRAIG}} \\ &= r_0 - U_n^{\text{NE}} J_n^{\text{NE}} y_n^{\text{CRAIG}} - e_n^{\text{T}} y_n^{\text{CRAIG}} \beta_{n+1}^{\text{NE}} u_{n+1}^{\text{NE}} \\ &= -\xi_n \beta_{n+1}^{\text{NE}} u_{n+1}^{\text{NE}}, \end{aligned}$$

with

$$\|r_n^{\text{CRAIG}}\|_2 = \beta_{n+1}^{\text{NE}} |\xi_n|.$$

Here we used again (6.8.8) and (6.8.9).

The next theorem summarizes the above derivations.

6.8 Implementations Based on Normal Equations

Theorem 6.8.3 Let v_n^{NE} (cf. (6.8.2)) denote the basis vectors generated by the bidiagonalization procedure (6.8.4) starting with $\beta_1^{NE} u_1^{NE} = r_0$. Furthermore, let $J_n^{NE}(\alpha_n^{NE}, \beta_n^{NE})$ (cf. (6.8.5)) be given.

(a) The nth CRAIG iterate is given by

$$x_n^{CRAIG} = x_{n-1}^{CRAIG} + \xi_n v_n^{NE}, \quad x_0^{CRAIG} = x_0,$$

where

$$\xi_n = -\beta_n^{NE} \xi_{n-1}/\alpha_n^{NE}, \quad \xi_1 = \beta_1^{NE}/\alpha_1^{NE}.$$

(b) The norm of the corresponding residual is given by

$$\|r_n^{CRAIG}\|_2 = \beta_{n+1}^{NE} |\xi_n|.$$

MATLAB Implementation of CRAIG

Note that the CRAIG implementation requires two matrix vector multiplications per step. Moreover, Au and Av can share one storage place. Altogether, we arrive at the following operation count for CRAIG.

Work per iteration and storage requirements of CRAIG:

- Matrix-vector products: matvec(A,u), matvec(A,v)   2
- dot products: norm(u), norm(v)   2
- additional flops: Au-beta*v, v/alpha,
  Av-alpha*u, u/beta,
  xCRAIG+xi*v   8N
- vectors to be stored: u, Au, v, xCRAIG   4

```
function [xCRAIG,norm_rCRAIG]=CRAIG(A,f,x0,n_max,tol)
%
% computes the solution of Ax=f using the CRAIG scheme.
%
% startvector:                          x0
% maximal number of iterations:         n_max
% desired accuracy of the residual:     tol
%
% uses: matvec
%

%%initialize
    n=1; N=length(x0);
    v=zeros(N,1);
    u=f-matvec(A,x0); beta=norm(u); u=u/beta;
    xi=-1;
    xCRAIG=x0; norm_rCRAIG=beta; norm_r0=beta;

    while (n < n_max+1) & (norm_rCRAIG/norm_r0 > tol)
        n=n+1;

    %%Bidiagonalization
        Au=matvec(A,u);
        v=Au-beta*v; alpha=norm(v); v=v/alpha;
        Av=matvec(A,v); beta_old=beta;
        u=Av-alpha*u; beta=norm(u); u=u/beta;

    %%update
        xi=-xi*beta_old/alpha;
        xCRAIG=xCRAIG+xi*v;

    %%compute norm
        norm_rCRAIG=beta*abs(xi);

    end %while
return;
```

6.9 Comparison of the Various Methods

In the last sections we discussed several parameter free methods that we believe are the most important for the solution of symmetric (indefinite) systems. Naturally, one might ask, do the introduced schemes differ significantly in their capabilities? If so, which of them is the method of choice? In this section we will report on some numerical experiments and partly answer these questions.

If the measure of the error in the iteration process is based on the 2-norm of the residual, then clearly the MR method has its advantages

$$\|r_n^{\mathrm{MR}}\|_2 \leq \|r_n^{\mathrm{OR}}\|_2.$$

6.9 Comparison of the Various Methods

Moreover, we know (cf. (6.8.12))

$$\|r_n^{\text{LSQR}}\|_2 \leq \|r_n^{\text{CRAIG}}\|_2.$$

One may ask, whether there is also some correspondence between the 2-norms of the respective errors. In fact, for the normal equations we have (cf. (6.8.14))

$$\|\varepsilon_n^{\text{CRAIG}}\|_2 \leq \|\varepsilon_n^{\text{LSQR}}\|_2.$$

For positive definite A it is known (cf. Hestenes and Stiefel [81, Theorem 7.5]) that

$$\|\varepsilon_n^{\text{OR}}\|_2 \leq \|\varepsilon_n^{\text{MR}}\|_2.$$

For indefinite systems, however, there is no such inequality. Here "anything is possible". This will be demonstrated in Example 6.9.2.

It should be mentioned that the statements of Paige and Saunders [95, p. 624] (" it is theoretically interesting to compare these approximations $(x_n^{\text{ME}}, x_n^{\text{OR}})$ to x_*") and of Szyld and Widlund [123, p. 14] (" x_n^{ME} is a better approximation than x_n^{OR} of x_* in the l_2-norm ") are somewhat misleading. A correct statement is (see the derivations for the SYMMLQ implementation)

$$x_n^{\text{OR}} = x_n^{\text{ME}} - s_n \hat{\eta}^{\text{ME}} w_{n+1}, \quad x_n^{\text{ME}} - x_0 \perp w_{n+1},$$

and consequently

$$\|x_n^{\text{ME}} - x_0\|_2 \leq \|x_n^{\text{OR}} - x_0\|_2.$$

From a numerical point of view, it is desirable to have monotonicity. All methods, but the Galerkin approach, fulfill this property

$$\|r_n^{\text{MR}}\|_2 \leq \|r_{n-1}^{\text{MR}}\|_2, \quad \|r_n^{\text{LSQR}}\|_2 \leq \|r_{n-1}^{\text{LSQR}}\|_2,$$
$$\|\varepsilon_n^{\text{ME}}\|_2 \leq \|\varepsilon_{n-1}^{\text{ME}}\|_2, \quad \|\varepsilon_n^{\text{CRAIG}}\|_2 \leq \|\varepsilon_{n-1}^{\text{CRAIG}}\|_2.$$

Of course, for positive definite A we have

$$\|\varepsilon_n^{\text{OR}}\|_A \leq \|\varepsilon_{n-1}^{\text{OR}}\|_A.$$

Before we discuss some numerical experiments, a few more general comments. In all our experiments we have **not** observed any significant difference in computing the

- OR iterates by SYMMLQ or by MINRES;
- MR iterates by MINRES or by SYMMLQ or by MCR;
- ME iterates by SYMMLQ or by STOD.

It would be interesting to see examples (if any exist), where these observations do not hold.

Finally, the following table provides a comparison of the operation count and the storage requirement for all discussed methods.

Method	matvec's	dot products	add. flops	storage
CG	1	2	6N	4
CR	1	2	8N	5
STOD	1	3	12N	8
MCR	1	3	12N	8
SYMMLQ	1	2	12N	6
MINRES	1	2	12N	7
LSQR	2	2	11N	5
CRAIG	2	2	8N	4

Table 6.9.1 Operation count and storage requirement.

The table clearly shows that CG and CR are the cheapest implementations. These methods, however, may break down and therefore one may use stable implementations like SYMMLQ and MINRES for the solution of indefinite systems. Note that STOD/MCR require one more dot product than SYMMLQ/MINRES. We would like to mention, that there is yet another possibility to stabilize the CG/CR implementations. The idea is to "skip" the critical situations by certain "look-ahead" strategies. These methods were investigated in great detail by Modersitzki [92].

Example 6.9.2 Let us now report on numerical results for the linear system $A(350, 25)x = f$, where $A(350, 25)$ is the Helmholtz matrix defined in (5.1.2). The right hand side f was chosen such that $x_* = (1, 1, \ldots, 1)^T$ is the solution. A random vector with components uniformly distributed in $[-1, 1]$ was selected as initial approximation x_0. In order to judge the obtained convergence rates, we also computed the minimal error $\|\varepsilon_n^{\text{MtE}}\|_2$ (cf. (6.7.1)) using MATHEMATICA with prec=150 decimal digits. The starting guess was produced by MATHEMATICA. All other computations

6.9 Comparison of the Various Methods

were carried out in MATLAB. The iterations were halted once the relative residual was brought below 10^{-10}

$$\|r_n\|_2/\|r_0\|_2 \leq 10^{-10}.$$

Figure 6.9.3 shows the convergence history for scaled residual norm. In order to keep the issues of interest clear, we computed the norm of the residuals by actually evaluating $\|f - Ax_n\|_2$ rather than by using one of the update formulae.

As the plot indicates, the convergence curve for MR is smooth. The plot also shows the typical oscillations in the OR convergence curve, and the correspondence between "peaks and plateaux" in the norm of the residuals of the OR and MR iterates. In this case, the ME iterates converge slowly. For this example, LSQR [CRAIG] converged in 570 [572] (!) steps.

Figure 6.9.4 shows the convergence behavior in terms of the scaled error norm. Observe that it is (in general) impossible to predict which method will produce the smallest error.

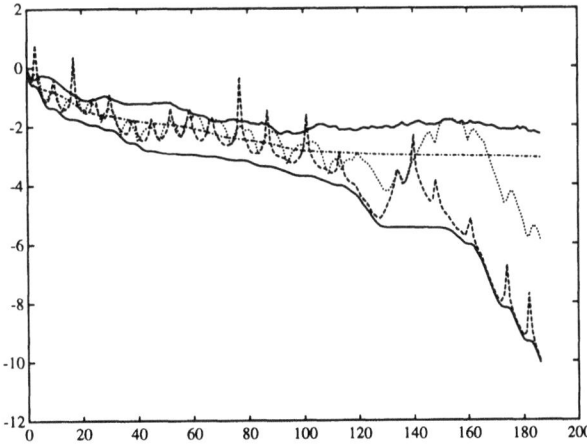

Figure 6.9.3 Residual norms versus the number of iterations for the matrix $A(350, 25)$. Lower solid curve: $\log_{10} \|f - Ax_n^{\mathrm{MR}}\|_2/\|r_0\|_2$; dashed curve: $\log_{10} \|f - Ax_n^{\mathrm{OR}}\|_2/\|r_0\|_2$; dotted curve: $\log_{10} \|f - Ax_n^{\mathrm{ME}}\|_2/\|r_0\|_2$; dashed-dotted curve: $\log_{10} \|f - Ax_n^{\mathrm{LSQR}}\|_2/\|r_0\|_2$; upper solid curve: $\log_{10} \|f - Ax_n^{\mathrm{CRAIG}}\|_2/\|r_0\|_2$.

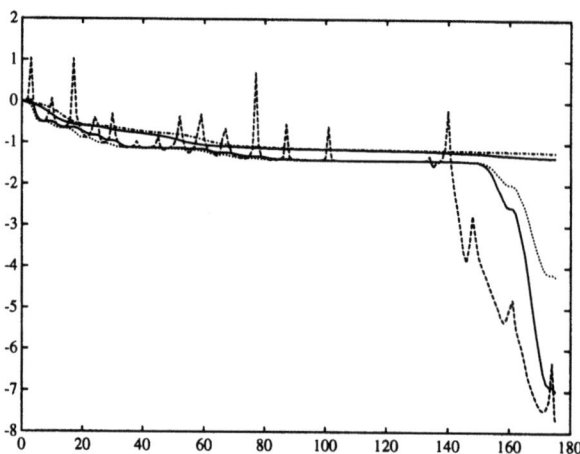

Figure 6.9.4 Error norms versus the number of iterations for the matrix $A(350,25)$. Lower solid curve: $\log_{10}\|x_* - x_n^{\mathrm{MR}}\|_2/\|\varepsilon_0\|_2$; dashed curve: $\log_{10}\|x_* - x_n^{\mathrm{OR}}\|_2/\|\varepsilon_0\|_2$; dotted curve: $\log_{10}\|x_* - x_n^{\mathrm{ME}}\|_2/\|\varepsilon_0\|_2$; dashed-dotted curve: $\log_{10}\|x_* - x_n^{\mathrm{LSQR}}\|_2/\|\varepsilon_0\|_2$; upper solid curve: $\log_{10}\|x_* - x_n^{\mathrm{CRAIG}}\|_2/\|\varepsilon_0\|_2$.

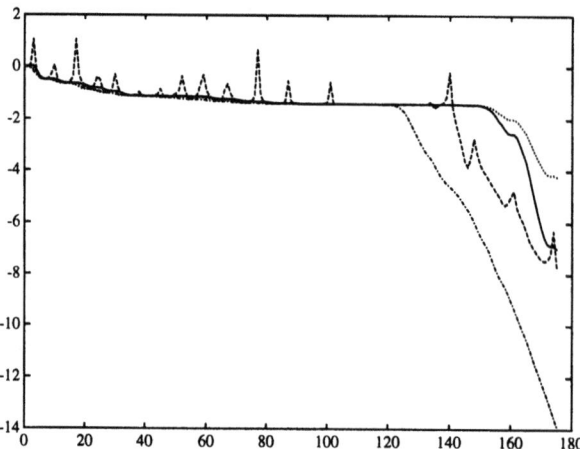

Figure 6.9.5 Error norms versus the number of iterations for the matrix $A(350,25)$. Solid curve: $\log_{10}\|x_* - x_n^{\mathrm{MR}}\|_2/\|\varepsilon_0\|_2$; dashed curve: $\log_{10}\|x_* - x_n^{\mathrm{OR}}\|_2/\|\varepsilon_0\|_2$; dotted curve: $\log_{10}\|x_* - x_n^{\mathrm{ME}}\|_2/\|\varepsilon_0\|_2$; dashed-dotted curve: $\log_{10}\|x_* - x_n^{\mathrm{MtE}}\|_2/\|\varepsilon_0\|_2$.

6.9 Comparison of the Various Methods

It is interesting to see how these curves compare to the very best error curve produced by the MtE iterates. Here, Figure 6.9.5 shows that, for this example, all curves have (at least) the same qualitative behavior. □

Example 6.9.2 suggests that the convergence behavior of the normal equation approach is pretty bad. This is indeed true for many matrices. Nevertheless there are special matrices where solving the normal equation is fine.

Symmetric Spectrum

The next example is of rather "academic" nature.

Example 6.9.6 Here A is a diagonal matrix, where the eigenvalues are uniformly distributed in two intervals which are symmetric with respect to the origin $[-a, -b] \cup [b, a]$. We have taken $N = 100$, $a = 2.0$, and $b = 0.5$. The right hand side, the solution, and the initial guess are chosen as described in Example 6.9.2.

The norm of the corresponding residuals is plotted in the next figure.

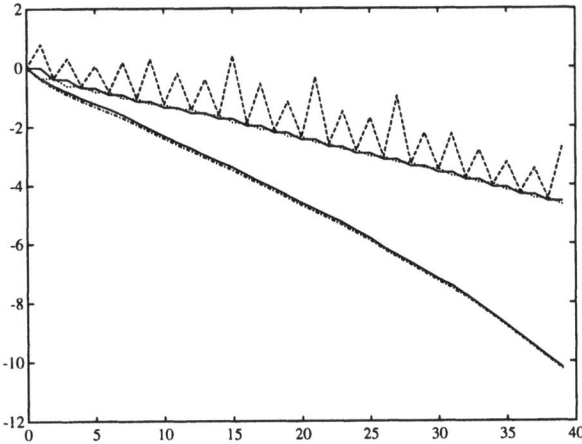

Figure 6.9.7 Residual norms versus the number of iterations for $N = 100$, $a = 2.0$, and $b = 0.5$. Upper solid curve: $\log_{10} \|f - Ax_n^{\text{MR}}\|_2/\|r_0\|_2$; dashed curve: $\log_{10} \|f - Ax_n^{\text{OR}}\|_2/\|r_0\|_2$; dashed-dotted curve: $\log_{10} \|f - Ax_n^{\text{LSQR}}\|_2/\|r_0\|_2$; dotted curve: $\log_{10} \|f - Ax_n^{\text{ME}}\|_2/\|r_0\|_2$; lower solid curve: $\log_{10} \|f - Ax_n^{\text{CRAIG}}\|_2/\|r_0\|_2$.

The convergence behavior of the various schemes for this particular example is well understood. The next lemma shows that the orthogonal polynomials with respect to the inner product (cf. 1.1.13)

$$\langle p, q \rangle_{\text{GAL}} = r_0^{\text{T}} p(A) q(A) r_0 = \sum_{j=1}^{L} \sigma_j^2 p(\lambda_j) q(\lambda_j).$$

inherit the symmetry of the eigenvalues.

Lemma 6.9.8 *Assume that the symmetric matrix A has a "symmetric" expansion into orthonormal eigenvectors*

$$\lambda_j = -\lambda_{L+1-j} + \tau,$$

$$r_0 = \sum_{j=1}^{L} \sigma_j z_j, \quad \sigma_j = \sigma_{L+1-j}, \quad j = 1, 2, \ldots, l, \ L = 2l,$$

for a real parameter $\tau \in \mathbb{R}$. Then the orthonormal polynomials with respect to $\langle \cdot, \cdot \rangle_{\text{GAL}}$ satisfy

$$\psi_{2n}(t) = q_n(t(t-\tau)) \quad \text{and} \quad \psi_{2n+1}(t) = (t - \tau/2) \hat{q}_n(t(t-\tau)),$$

where q_n and $\hat{q}_n$ are polynomials of exact degree n, respectively.

Proof. Consider the linear mapping $\varphi(t) := -t + \tau$ and let ψ_k^{MO} denote the monic orthogonal polynomials with respect to $\langle \cdot, \cdot \rangle_{\text{GAL}}$. By assumption we have

$$\|\psi_k^{\text{MO}}(t)\|_{\text{GAL}}^2 = \sum_{j=1}^{L} \sigma_j^2 \left(\psi_k^{\text{MO}}(\lambda_j) \right)^2$$

$$= \sum_{j=1}^{L} \sigma_j^2 \left(\psi_k^{\text{MO}}(\varphi(\lambda_j)) \right)^2$$

$$= \|\psi_k^{\text{MO}}(\varphi(t))\|_{\text{GAL}}^2.$$

As a direct consequence of Theorem 2.1.4(b) we obtain

$$\psi_k^{\text{MO}}(t) = (-1)^n \psi_k^{\text{MO}}(\varphi(t)).$$

6.9 Comparison of the Various Methods

Hence the zeros of ψ_{2n}^{MO} come in pairs

$$\psi_{2n}^{MO}(t) = \prod_{j=1}^{n}(t-x_j)(t-\varphi(x_j)) = \prod_{j=1}^{n}(t(t-\tau) - x_j(x_j-\tau)) = q_n^{MO}(t(t-\tau)).$$

The proof of the odd case makes in addition use of the observation that $\varphi(x) = x$ if, and only if $x = \tau/2$. □

The above numerical example corresponds to the special case $\tau = 0$.

Theorem 6.9.9 *Assume that the symmetric matrix A has a "symmetric" expansion into orthonormal eigenvectors*

$$\lambda_j = -\lambda_{L+1-j},$$
$$r_0 = \sum_{j=1}^{L} \sigma_j z_j, \quad \sigma_j = \sigma_{L+1-j}, \quad j = 1, 2, \ldots, l, \; L = 2l.$$

Then we have

(a) $\quad x_{2n+1}^{MR} = x_{2n}^{MR} = x_n^{LSQR}, \quad n = 0, 1, \ldots;$

(b) $\quad x_{2n}^{ME} = x_{2n-1}^{ME} = x_{2n}^{OR} = x_n^{CRAIG}, \quad n = 1, 2, \ldots.$

Moreover, no odd OR iterate x_{2n-1}^{OR} exists.

Proof. To show (a) we make use of Theorem 2.5.1(a), Lemma 6.9.8, and (6.8.12)

$$\psi_{2n+1}^{MR}(t) = \psi_{2n}^{MR}(t) = \frac{\sum_{j=0}^{2n+1} \psi_j(t)\psi_j(0)}{\sum_{j=0}^{2n+1} \psi_j(0)^2} = \frac{\sum_{j=0}^{n} q_j(t^2)q_j(0)}{\sum_{j=0}^{n} q_j(0)^2} =: P_n(t^2),$$

with
$$\|P_n(A^2)r_0\|_2 = \min\left\{\|p(A^2)r_0\|_2 : p \in \Pi_n, \; p(0) = 1\right\}.$$

For the proof of (b) observe that by Lemma 6.9.8

$$\psi_{2n}'(t) = 2tq_n'(t^2) \quad \text{and} \quad \psi_{2n+1}'(t) = 2t^2\hat{q}_n'(t^2) + \hat{q}_n(t^2).$$

Hence the Christoffel Darboux formula (2.6.12) simplifies

$$\psi_{2n}^{\mathrm{ME}}(t) = \frac{\psi'_{2n}(0)\psi_{2n+1}(t) - \psi'_{2n+1}(0)\psi_{2n}(t)}{\psi'_{2n}(0)\psi_{2n+1}(0) - \psi'_{2n+1}(0)\psi_{2n}(0)} = \frac{\psi_{2n}(t)}{\psi_{2n}(0)} = \psi_{2n}^{\mathrm{OR}}(t)$$

$$\psi_{2n-1}^{\mathrm{ME}}(t) = \frac{\psi'_{2n-1}(0)\psi_{2n}(t) - \psi'_{2n}(0)\psi_{2n-1}(t)}{\psi'_{2n-1}(0)\psi_{2n}(0) - \psi'_{2n}(0)\psi_{2n-1}(0)} = \frac{\psi_{2n}(t)}{\psi_{2n}(0)} = \psi_{2n}^{\mathrm{OR}}(t),$$

which proves the first two identities. As for (a), the third equality follows from Theorem 2.6.1 and (6.8.14). Finally, we learn from Lemma 6.9.8 that $\psi_{2n-1}(0) = 0$ which says that no odd OR iterate exists. □

We remark that an alternative proof may be found in Freund [47] (see also Freund, Golub, and Nachtigal [52, §4] and Fischer, Ramage, Silvester, and Wathen [45]).

This time the corresponding error curves look like

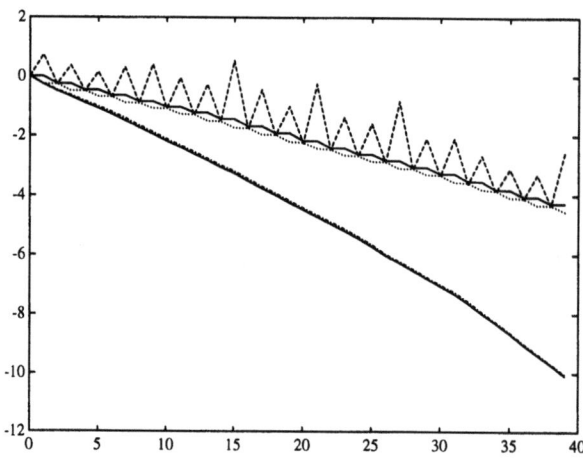

Figure 6.9.10 Error norms versus the number of iterations for $N = 100$, $a = 2.0$, and $b = 0.5$. Upper solid curve: $\log_{10} \|x_* - x_n^{\mathrm{MR}}\|_2 / \|\varepsilon_0\|_2$; dashed curve: $\log_{10} \|x_* - x_n^{\mathrm{OR}}\|_2 / \|\varepsilon_0\|_2$; dotted curve: $\log_{10} \|x_* - x_n^{\mathrm{ME}}\|_2 / \|\varepsilon_0\|_2$; dashed-dotted curve: $\log_{10} \|x_* - x_n^{\mathrm{LSQR}}\|_2 / \|\varepsilon_0\|_2$; Lower solid curve: $\log_{10} \|x_* - x_n^{\mathrm{CRAIG}}\|_2 / \|\varepsilon_0\|_2$.

6.9 Comparison of the Various Methods

In other words, in exact arithmetic LSQR [CRAIG] produces with the same number of matrix-vector products and half the number of dot products the same residuals as MR [ME]. Dot products constitute a bottleneck on modern architectures. Therefore, on a parallel machine, it may pay off to use one of the normal equation solvers for matrices discussed in this subsection. □

7 Parameter Dependent Methods

In this chapter we are concerned with so-called parameter dependent methods. These are polynomial based iteration methods where the residual polynomial is generated beforehand based on some information about the underlying system.

If the three-term recurrence coefficients of the residual polynomial are known one may implement the resulting scheme with the help of Algorithm 6.2.1. Recall, that this procedure is extremely cheap in terms of arithmetic operations. In particular, it does not involve the computation of dot products, which constitutes a bottleneck on modern architectures.

We will discuss two particular parameter dependent schemes. Both methods require some a priori information on the spectrum of the given matrix A.

The first method generalizes the classical Chebyshev iteration (cf. Golub and Varga [65], Eiermann, Niethammer, and Varga [35]) for positive definite systems to the case of symmetric indefinite systems. Here, the required input is just a compact subset Ω of the real axis known to contain all eigenvalues of A, i.e., $\sigma(A) \subset \Omega$. The residual polynomial is the optimal polynomial with respect to Ω.

The second method requires an approximation to the eigenvalue distribution of the given matrix. Here, the residual polynomials are the scaled orthogonal polynomials with respect to the estimated distribution function. One may view this approach as a generalization of a paper by Saad [107].

7.1 The Chebyshev Iteration for Symmetric Indefinite Systems

In this section we discuss some implementation details for the Chebyshev iteration when applied to symmetric indefinite systems $Ax = f$.

Let $E = [a, b] \cup [c, d]$, $a < b < 0 < c < d$ denote a set that contains all eigenvalues of A. Then the *Chebyshev iteration*

$$(7.1.1) \qquad r_n^{\text{CI}} = \mathcal{P}_n(A; E, 0) r_0$$

7.1 The Chebyshev Iteration for Symmetric Indefinite Systems

is defined by the optimal polynomial with respect to E (cf. (3.1.3))

(7.1.2) $\qquad \|\mathcal{P}_n(t; E, 0)\|_E = \min\{\|p\|_E : p \in \Pi_n, p(0) = 1\}.$

For the rest of this section we drop the dependence on E and 0, i.e., $\mathcal{P}_n(t) = \mathcal{P}_n(t; E, 0)$. Recall from Theorem 3.3.5(d) that $\mathcal{P}_n(t) = \mathcal{T}_n(t)/\mathcal{T}_n(0)$ is explicitly known when the associated Chebyshev polynomial $\mathcal{T}_n(t) = \mathcal{T}_n(t; E)$ has $n+2$ extremal points on E, which may be seen as a requirement for the set E. Sets with this property are characterized in Theorem 3.3.18. Let us briefly review the characterization. First, one maps E onto the unit interval (cf. (3.2.1))

$$l(E) = [-1, \hat{b}] \cup [\hat{c}, 1].$$

Then, the interior boundary points are expressed in terms of elliptic functions (cf. (3.3.13) and (3.3.14))

(7.1.3) $\qquad \hat{b} = 1 - 2\operatorname{sn}^2(\rho; k), \quad \text{and} \quad \hat{c} = 2\operatorname{sn}^2(K + \rho; k) - 1,$

where the modulus k is given by

$$k^2 := \frac{2(\hat{c} - \hat{b})}{(1 - \hat{b})(1 + \hat{c})}$$

and the number ρ is defined implicitly by either one of the equations in (7.1.3). Here, $K = K(k)$ denotes the complete elliptic integral of the first kind (cf. (3.3.9)). Now Theorem 3.3.18 states that $\mathcal{T}_n$ has $n+2$ extremal points on E, if

(7.1.4) $\qquad \rho = \rho_n = -\frac{m}{n} K(k), \quad m \in \{1, 2, \ldots, n-1\},$

is essentially a rational number.

Furthermore, $\mathcal{T}_n$ fulfills a three-term recurrence relation (see Theorem 3.3.7). Based on these coefficients, the Chebyshev iteration may be implemented with the help of Algorithm 6.2.1. Here, the coefficients of the residual polynomials may be computed according to Lemma 2.4.3.

To give an idea of the performance of the Chebyshev iteration, let us start with a contrived example.

Example 7.1.1 Here, we have chosen

$$n = 7, \ m = 4, \ k = 0.9 \ \text{and} \ \rho_n = -\frac{m}{n} K(k).$$

With this settings we then computed $\hat{b}$ and $\hat{c}$ according to (7.1.3), which results in

$$\hat{b} = -0.5723\ldots \quad \text{and} \quad \hat{c} = 0.1776\ldots$$

Finally, we constructed a diagonal matrix with 100 eigenvalues equidistantly distributed in $[-1,\hat{b}] \cup [\hat{c},1]$. Again, the right hand side f was chosen such that $x_* = (1,1,\ldots,1)^T$ is the solution. And a random vector x_0 with elements uniformly distributed in $[-1,1]$ served as the initial guess.

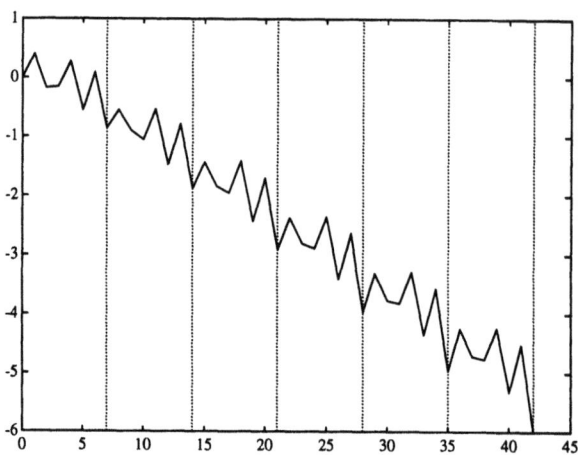

Figure 7.1.2 Residual norms versus the number of iterations. Solid curve: $\log_{10}\|f - Ax_n^{\mathrm{CI}}\|_2/\|r_0\|_2$; vertical dotted lines: $n \equiv j\cdot 7$, $j = 1,2,\ldots$

We learn from Figure 7.1.2 that the slope of the convergence curve is fine, though the curve itself behaves slightly irregular. This may be explained as follows. By construction and by Corollary 3.3.6 we know that the residual polynomial $\mathcal{P}_k$ is optimal, in the sense of (7.1.2), whenever the index k is a multiple of $n = 7$. This is indicated by the dotted lines in the plot. For degrees $j\cdot n < k < (j+1)\cdot n$ the residual polynomials $\mathcal{P}_k$, in general, do not fulfill any optimality condition. □

However, in view of Figure 7.1.2, it seems to be natural to look for scheme which exclusively compute the iterates x_n associated with optimal polynomials $\mathcal{P}_n$. In fact, it is possible to devise a scheme which "jumps over" unwanted iterates. To describe the method, let $\mathcal{T}_n$ denote a Chebyshev

7.1 The Chebyshev Iteration for Symmetric Indefinite Systems

polynomial with $n+2$ extremal points and let $L_n = \|T_n\|_E$. We know from Corollary 3.3.6 that the residual polynomial

$$\mathcal{P}_{n \cdot l}(t) = \frac{T_l(T_n(t)/L_n)}{T_l(T_n(0)/L_n)}$$

is optimal for any $l \in \mathbb{N}$. Here, T_l denotes the classical Chebyshev polynomial for the unit interval. Hence, the polynomial based iteration method defined by

(7.1.5) $$\hat{r}_l^{\mathrm{CI}} = \mathcal{P}_{n \cdot l}(A) r_0, \quad l = 1, 2, \ldots,$$

has the wanted property. We call the resulting scheme the *Leapfrog variant of the Chebyshev iteration*. With the settings

$$\hat{A} := T_n(A)/L_n \quad \text{and} \quad \hat{\xi} := T_n(0)/L_n$$

equation (7.1.5) becomes

$$\hat{r}_l^{\mathrm{CI}} = \frac{T_l(\hat{A})}{T_l(\hat{\xi})} r_0$$

Consequently, one can take advantage of the three-term recurrence relation of the T_l's to work out formulae for the computation of the iterates x_l^{CI}. We omit these straightforward calculations. Nevertheless, two little remarks are appropriate here. Each iteration step involves the computation of $\hat{A}y$, for some vector y. The cost of this operation is n matrix vector multiplies. Observe, that one may turn the operation $\hat{A}y$ to advantage once the underlying matrix A has been loaded into "high speed memory". Furthermore, one has to compute the norm of L_n. To perform this calculation one could make use of the explicit representation of L_n in Theorem 3.3.18, which involves the Jacobian Theta function. An alternate method is provided by Lemma 3.3.3(a). It states that all four boundary points are extremal points which, for example, implies that $L_n = T_n(1)$.

We finish this section with a more sophisticated example.

Example 7.1.3 This time we will apply the Chebyshev iteration to the Helmholtz matrix $A(350, 25)$ (cf. (5.1.2)) of order $N = 625$. The right hand side f was chosen such that $x_* = (1, 1, \ldots, 1)^{\mathrm{T}}$ is the solution. And a random vector x_0 with elements uniformly distributed in $[-1, 1]$ served as the initial guess.

Recall from Example 5.1.1 that the eigenvalues of $A(350, 25)$ are located in $E = [-.489, -.00019] \cup [0.0433, 7.46]$. We remark that the asymptotic convergence rate associated with this set is $\kappa(E) = 0.9985\ldots$

For the numerical test, however, we estimated the spectrum. To this end, we applied 20 steps of the Lanczos method to $A(350, 25)$ starting with r_0. Then we computed the Ritz values and the harmonic Ritz values and applied the "intersection technique" (5.2.1). This produced the following approximation $\hat{E} \subset E$

$$\hat{E} = [-.482, -.328] \cup [.482, 7.41].$$

Note that we missed some eigenvalues around the origin. The next step was to compute the optimal polynomial with respect to $\hat{E}$.

This is possible (cf. (7.1.3) and (7.1.4)) if the number ρ associated with $\tilde{E}$ is "rational" $\rho/K = m/n$. So, using the algorithm described in Section 3.4 we computed ρ/K, which (of course) turned out to be not rational. Therefore we approximated ρ/K by a rational number. Here, we took $n = 13$ and $m = 11$ which are connected to the set

$$\tilde{E}_n = [-.482, -.322] \cup [.518, 7.41].$$

Notice that we kept the outer boundary points, since these are in general good approximations to the extreme eigenvalues. Finally, we computed the residual polynomials $\mathcal{P}_n(t; \tilde{E}_n, 0)$ associated with $\tilde{E}_n$ and performed the Chebyshev iteration

$$r_n^{\text{CI}} = \mathcal{P}_n(A(350, 25); \tilde{E}_n, 0)r_0.$$

Altogether, we arrive at the following convergence history. One should keep in mind, that by construction $\mathcal{P}_k(t; \tilde{E}_n, 0)$ is optimal whenever k is a multiple of $n = 13$. To set the obtained results into perspective, we computed in addition the MR iterates x_n^{MR} by the MINRES scheme.

So, the scheme starts to converge nicely and levels out after a while. This is partly due to the fact that $\tilde{E}_n$ does not correctly approximate the spectrum. Also, since the residual polynomials for the Chebyshev iteration are computed beforehand, they do not adjust during the iteration to the underlying eigenvalue distribution, in contrast to parameter free methods.

7.2 Methods Based on the Eigenvalue Distribution

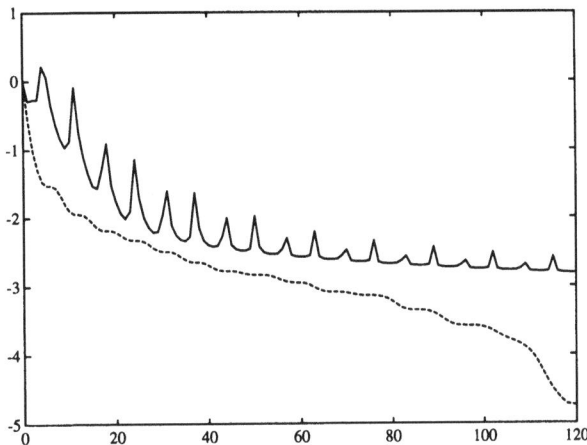

Figure 7.1.4 Residual norms versus the number of iterations. Solid curve: $\log_{10} \|f - Ax_n^{\text{CI}}\|_2/\|r_0\|_2$; dashed curve: $\log_{10} \|f - Ax_n^{\text{MR}}\|_2/\|r_0\|_2$.

A common way to overcome these problems is to set up a hybrid scheme. That is, one essentially mixes the Chebyshev iteration with a parameter free method, say a MR-scheme. The idea is that the MR-scheme should acquire information about the given matrix A while the successively applied Chebyshev iteration, designed with respect to the obtained information, should save inner products. However, to find the correct switching criteria between the two methods is quite a tricky business and will not be discussed here. □

7.2 Methods Based on the Eigenvalue Distribution

In this section we will devise parameter dependent methods for the solution of
$$Ax = f, \quad A \in \mathbb{R}^{N \times N},$$
which are based on some knowledge of the eigenvalue distribution of A. Before we start, we like to point out, that the material presented in this section is the outcome of joint work with Roland Freund. The work is still in progress (compare Fischer and Freund [38]).

Most of the presented parameter free methods are based either on the inner product (cf. Section 6.1, Gal approach)

(7.2.1) $$\langle p, q \rangle_{\text{GAL}} = r_0^T p(A) q(A) r_0 = \int_a^d p(t) q(t) d\sigma(t)$$

or on the inner product (cf. Section 6.1, MR approach)

$$(7.2.2) \qquad \langle p, tq \rangle_{\text{GAL}} = r_0^T p(A) A q(A) r_0 = \int_a^d p(t)q(t)t d\sigma(t).$$

Recall that each of these methods has to perform two dot products per iteration step. Here, a dot product is a vector-vector multiplication $x^T y$, $x, y \in \mathbb{R}^N$. The required dot products are nothing but the evaluation of (7.2.1) or (7.2.2).

Now, the idea is to approximate the inner products (7.2.1) and (7.2.2) by new inner products which can be evaluated without dot products. We proceed as follows. First, we approximate the eigenvalue distribution of A by the method described in Section 5.3. This yields a monotone piecewise cubic interpolant s satisfying

$$s(t_j) = \vartheta_j, \quad t_0 < t_1 < \cdots < t_{m+1}.$$

The function s gives rise to the new inner product (cf. 5.3.5)

$$\langle p, q \rangle_s = \int_{t_0}^{t_{m+1}} p(t)q(t) ds(t) = \int_{t_0}^{t_{m+1}} p(t)q(t) s'(t) dt.$$

Having computed the approximation s one may now generate (beforehand) the orthogonal polynomials ψ_n^s with respect to $\langle \cdot, \cdot \rangle_s$, i.e.,

$$(7.2.3) \qquad \langle \psi_k^s, \psi_l^s \rangle_s = 0, \quad \text{for} \quad k \neq l,$$

or, likewise, the orthogonal polynomials $\hat{\psi}_n^s$ with respect to $\langle \cdot, t \cdot \rangle_s$, i.e.,

$$(7.2.4) \qquad \langle \hat{\psi}_k^s, t \hat{\psi}_l^s \rangle_s = 0, \quad \text{for} \quad k \neq l.$$

Of course, one could as well generate Hermite kernel polynomials with respect to $\langle \cdot, \cdot \rangle_s$. Here we focus on orthogonal polynomials (7.2.3) and on kernel polynomials (7.2.4).

In view of Cor. 2.4.7 and Theorem 2.5.1(c) the associated residual polynomials are characterized as the solution of an approximation problem. We have

$$\frac{1}{(\psi_n^s(0))^2} \langle \psi_n^s, t^{-1} \psi_n^s \rangle_s = \min\{\langle p, t^{-1} p \rangle_s : p \in \Pi_n, p(0) = 1\},$$

7.2 Methods Based on the Eigenvalue Distribution

if the associated Jacobi matrix J_n^s is positive definite, and

$$(7.2.5) \qquad \frac{1}{(\hat{\psi}_n^s(0))^2} \langle \hat{\psi}_n^s, \hat{\psi}_n^s \rangle_s = \min\{\langle p, p \rangle_s : p \in \Pi_n, p(0) = 1\}.$$

Both sets of orthogonal polynomials may be used to define a polynomial based iteration method by

$$(7.2.6) \qquad r_n^s = \frac{\psi_n^s(A)}{\psi_n^s(0)} r_0,$$

or by

$$(7.2.7) \qquad \hat{r}_n^s = \frac{\hat{\psi}_n^s(A)}{\hat{\psi}_n^s(0)} r_0,$$

respectively. These methods share the properties with the corresponding methods discussed in the chapter on parameter free methods, but with respect to a different inner product. For example, let us define the induced norm (cf. Section 1.1)

$$(7.2.8) \qquad \|u\|_s := \sqrt{\langle p^{(u)}, p^{(u)} \rangle_s}, \quad \text{where} \quad u = p^{(u)}(A) r_0.$$

With this norm we may rewrite the variational property (7.2.5) as

$$\|\hat{r}_n^s\|_s = \min\{\|f - Ax_n\|_s : x \in x_0 + \mathcal{K}_n(A, r_0)\}.$$

In other words, (7.2.7) defines a scheme with minimal residual with respect to $\|\cdot\|_s$.

The next examples give an idea about the performance of methods based on the inner product $\langle \cdot, \cdot \rangle_s$. Here, we considered the Helmholtz matrix $A(50, 25)$ (cf. (5.1.2)) of order $N = 625$. The right hand side f was chosen such that $x_* = (1, 1, \ldots, 1)^T$ is the solution. And a random vector x_0 with elements uniformly distributed in $[-1, 1]$ served as the initial guess. To generate the interpolation data for the computation of the of the monotone piecewise cubic interpolant $s^{(k)}$ we first applied $k = 10, 20, 40$ steps of the Lanczos process to $A(50, 25)$ starting with r_0. In order to keep the issues of interest clear, we did not make further use of the obtained Lanczos vectors.

Example 7.2.1 For this example we just computed the three-term recurrence coefficients of the orthogonal residual polynomials $\psi_n^{\rm s}(t)/\psi_n^{\rm s}(0)$ (cf. 7.2.3) and $\hat{\psi}_n^{\rm s}(t)/\hat{\psi}_n^{\rm s}(0)$ (cf. 7.2.4), respectively. Then we implemented the resulting polynomial based methods with the help of Algorithm 6.2.1. One may view them as CG-type and CR-type implementations with respect to $\langle\cdot,\cdot\rangle_{\rm s}$, which in particular implies that a breakdown is possible. Based on this relationship we denote the iterates associated with the inner product $\langle\cdot,\cdot\rangle_{\rm s}$ by $x_n^{\rm CGs}$ and $x_n^{\rm CRs}$, respectively. The performance, with respect to $s^{(k)}, k = 10, 20, 40$, is illustrated in the next three figures. To set the obtained convergence rates into perspective we plotted in addition the convergence curve for the iterates produced by MINRES.

It is apparent that the "CRs iteration" benefits from the improved approximation of the eigenvalue distribution, whereas the "CGs iteration" seems not to benefit from the increased number of sample points. Anyway, both approaches need further investigations to fully understand their behavior.

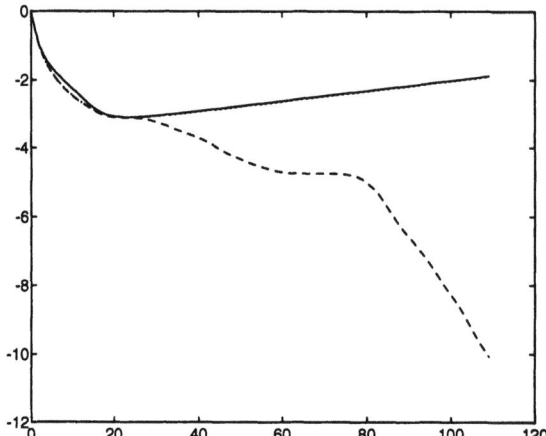

Figure 7.2.2 True Residual norms versus the number of iterations for the matrix $A(50, 25)$, the inner product is based on $s^{(10)}$. Solid curve: $\log_{10} \|f - Ax_n^{\rm CGs}\|_2/\|r_0\|_2$; dashed curve: $\log_{10} \|f - Ax_n^{\rm MR}\|_2/\|r_0\|_2$; dotted curve: $\log_{10} \|f - Ax_n^{\rm CRs}\|_2/\|r_0\|_2$.

7.2 Methods Based on the Eigenvalue Distribution

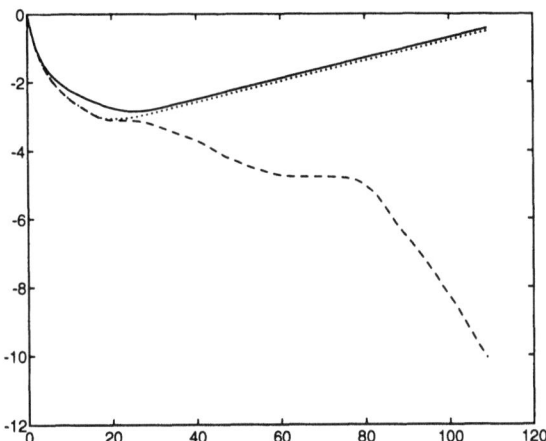

Figure 7.2.3 True Residual norms versus the number of iterations for the matrix $A(50,25)$, the inner product is based on $s^{(20)}$. Solid curve: $\log_{10} \|f - Ax_n^{\text{CGs}}\|_2/\|r_0\|_2$; dashed curve: $\log_{10} \|f - Ax_n^{\text{MR}}\|_2/\|r_0\|_2$; dotted curve: $\log_{10} \|f - Ax_n^{\text{CRs}}\|_2/\|r_0\|_2$.

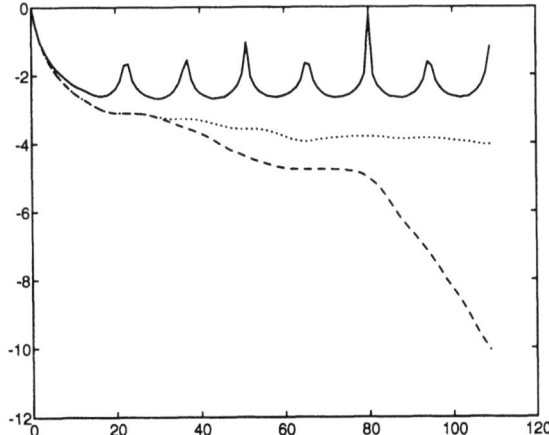

Figure 7.2.4 True Residual norms versus the number of iterations for the matrix $A(50,25)$, the inner product is based on $s^{(40)}$. Solid curve: $\log_{10} \|f - Ax_n^{\text{CGs}}\|_2/\|r_0\|_2$; dashed curve: $\log_{10} \|f - Ax_n^{\text{MR}}\|_2/\|r_0\|_2$; dotted curve: $\log_{10} \|f - Ax_n^{\text{CRs}}\|_2/\|r_0\|_2$.

□

Example 7.2.5 In the example above, we compared the Euclidian norm of the various residuals. This is not "fair", since the schemes based on the interpolant s are not aiming to minimize the Euclidian norm, whereas MINRES does. Therefore, it interesting to have a look at the "s-norm" (cf. (7.2.8)) of the residuals.

In this example, we implemented a MINRES-type approach. That is, we substituted the "Lanczos part" in MINRES by Algorithm 4.1.1 based on $\langle \cdot, \cdot \rangle_s$. Note, that we "automatically" obtain the s-norm of the updated residual $\|r_n^{\text{MRs}}\|_s$ in terms of the Givens parameters of the associated QR factorization (cf. Cor. 2.5.3).

So, except for the last experiment ($k = 40$) the Euclidian norm and the s-norm are quite different. Again, it is an open question what conclusions can be drawn from this observations.

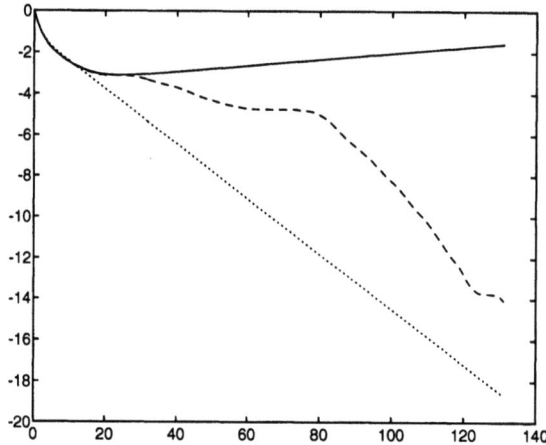

Figure 7.2.6 True Residual norms versus the number of iterations for the matrix $A(50, 25)$, the inner product is based on $s^{(10)}$. Solid curve: $\log_{10} \|f - Ax_n^{\text{MRs}}\|_2 / \|r_0\|_2$; dashed curve: $\log_{10} \|f - Ax_n^{\text{MR}}\|_2 / \|r_0\|_2$; dotted curve: $\log_{10} \|f - Ax_n^{\text{MRs}}\|_s / \|r_0\|_s$.

7.2 Methods Based on the Eigenvalue Distribution

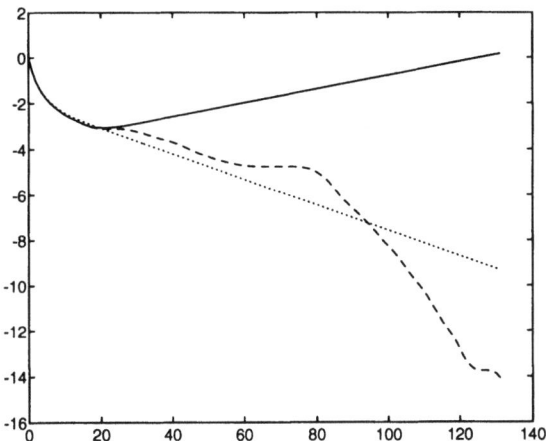

Figure 7.2.7 True Residual norms versus the number of iterations for the matrix $A(50, 25)$, the inner product is based on $s^{(20)}$. Solid curve: $\log_{10} \|f - Ax_n^{\mathrm{MRs}}\|_2/\|r_0\|_2$; dashed curve: $\log_{10} \|f - Ax_n^{\mathrm{MR}}\|_2/\|r_0\|_2$; dotted curve: $\log_{10} \|f - Ax_n^{\mathrm{MRs}}\|_{\mathrm{S}}/\|r_0\|_{\mathrm{S}}$.

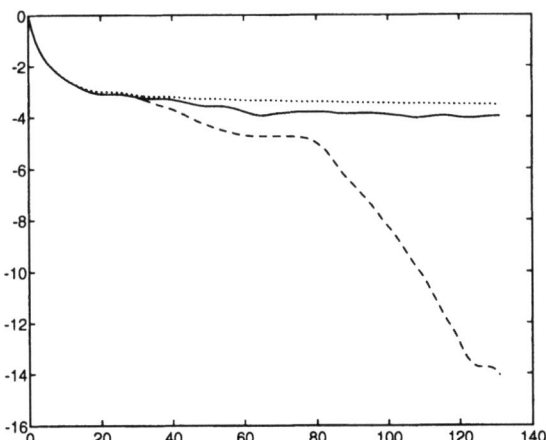

Figure 7.2.8 True Residual norms versus the number of iterations for the matrix $A(50, 25)$, the inner product is based on $s^{(40)}$. Solid curve: $\log_{10} \|f - Ax_n^{\mathrm{MRs}}\|_2/\|r_0\|_2$; dashed curve: $\log_{10} \|f - Ax_n^{\mathrm{MR}}\|_2/\|r_0\|_2$; dotted curve: $\log_{10} \|f - Ax_n^{\mathrm{MRs}}\|_{\mathrm{S}}/\|r_0\|_{\mathrm{S}}$.

□

8 The Stokes Problem

The purpose of this chapter is to introduce one of the "most prominent" sources of symmetric but indefinite linear systems. It stems from finite element (or finite volumes, finite differences, ...) discretizations of the Stokes equations. In a more general framework, the standard variational formulation of the Stokes equations turns out to be nothing but a saddle point problem. We will see that appropriate discretizations of such a problem lead to indefinite linear systems. It is precisely this problem class which motivated the search for stable and efficient solvers for symmetric but indefinite systems.

8.1 The Continuous Problem

The classical Stokes equations describe slow viscous incompressible flow in a flow domain $\Omega \subset \mathbb{R}^d$, $d = 2, 3$. To simplify the discussion (and notation), we focus on the case $d = 2$. The three dimensional case works along the same lines. Moreover, in order to avoid technical difficulties, we assume that the boundary $\Gamma = \partial \Omega$ is sufficiently smooth, e.g., Lipschitz continuous. More precisely we assume that Ω is a *polygonal domain*, i.e., the boundary Γ is polygonal curve and $\overline{\Omega} = \Omega \cup \Gamma$.

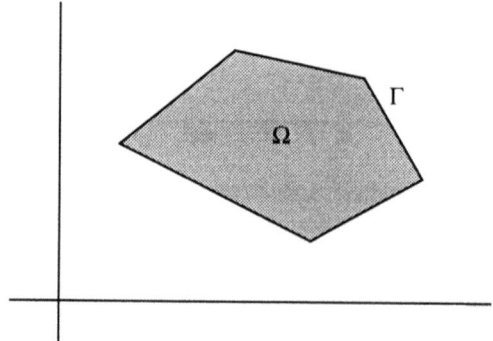

Figure 8.1.1 Flow domain Ω.

8.1 The Continuous Problem

There are several books dealing with the subject in great detail, including Brezzi and Fortin [12], Girault and Raviart [64], Gunzburger [74], and Johnson [84]. Here, we mainly followed Braess [10] and Hackbusch [76] (see also [77]).

Hilbert Spaces

In this subsection we briefly introduce some function spaces and collect some of their properties.

As usual, let $C^k(\Omega)$ denote the set of all k-times continuously differentiable functions $u : \Omega \to \mathbb{R}$, where $0 \leq k \leq \infty$. An important subset of $C^\infty(\Omega)$ is the space

$$C_0^\infty(\Omega) := \{u \in C^\infty(\Omega) : \mathrm{supp}(u) \text{ compact}, \mathrm{supp}(u) \subset \Omega, \mathrm{supp}(u) \cap \Gamma = \emptyset\}$$

of indefinitely differentiable functions with compact support in Ω. Furthermore, let

$$(8.1.1) \qquad L_2(\Omega) := \{q : \Omega \to \mathbb{R} : \int_\Omega q^2 \, d(x,y) < \infty\}$$

denote the set of all measurable functions whose squares are Lebesque integrable. Equipped with the inner product

$$(8.1.2) \qquad \langle p, q \rangle_{L_2} := \int_\Omega p \, q \, d(x,y), \quad \|q\|_{L_2} := \sqrt{\langle q, q \rangle_{L_2}},$$

the space $L_2(\Omega)$ is a Hilbert space.

Next we introduce the *Sobolev space*

$$(8.1.3) \qquad H^1(\Omega) := \{u \in L_2(\Omega) : \frac{\partial u}{\partial x}, \frac{\partial u}{\partial y} \in L_2(\Omega)\}$$

which is a Hilbert space with respect to

$$(8.1.4) \qquad \langle u, v \rangle_{H^1} := \int_\Omega u \, v + (\mathrm{grad}\, u)^T \mathrm{grad}\, v \, d(x,y), \quad \|u\|_{H^1} := \sqrt{\langle u, u \rangle_{H^1}}.$$

The closure of $C_0^\infty(\Omega)$ with respect to $\|\cdot\|_{H^1}$ is denoted by $H_0^1(\Omega)$. We remark that for sufficiently smooth boundary, e.g., Γ is Lipschitz continuous, the following characterization holds

$$H_0^1(\Omega) = \{u \in H^1(\Omega) : u(x,y) = 0 \text{ on } \Gamma\}.$$

The above concept generalizes to higher order derivatives, e.g.,

$$H^2(\Omega) := \{u \in L_2(\Omega) : \frac{\partial u^{(k_1,k_2)}}{\partial x^{k_1} \partial y^{k_2}} \in L_2(\Omega), \ k_i \in \{0,1,2\}, \ k_1 + k_2 \leq 2\},$$

and, for sufficiently smooth boundary Γ,

$$H_0^2(\Omega) = \{u \in H^2(\Omega) : u(x,y) = \frac{\partial u}{\partial n}(x,y) = 0 \text{ on } \Gamma\}.$$

The Continuous Stokes Problem

In this section we define the classical Stokes equations. To this end, let $f = [f_1 \ f_2]^T \in L_2(\Omega) \times L_2(\Omega)$ be given. Then the *Stokes problem* is:

Find a velocity $u = [u_1 \ u_2]^T : \Omega \to \mathbb{R}^2$ and a pressure $p : \Omega \to \mathbb{R}$ satisfying

(8.1.5)
$$\left. \begin{array}{r} -\left(\dfrac{\partial^2 u_1}{\partial^2 x} + \dfrac{\partial^2 u_1}{\partial^2 y}\right) + \dfrac{\partial p}{\partial x} = f_1 \\[2mm] -\left(\dfrac{\partial^2 u_2}{\partial^2 x} + \dfrac{\partial^2 u_2}{\partial^2 y}\right) + \dfrac{\partial p}{\partial y} = f_2 \\[2mm] \dfrac{\partial u_1}{\partial x} + \dfrac{\partial u_2}{\partial y} = 0 \end{array} \right\} \text{ in } \Omega.$$

Various kinds of boundary conditions may be used in connection with (8.1.5). For simplicity, we consider only the homogeneous no-flow boundary conditions

(8.1.6) $$u_1 = u_2 = 0 \quad \text{on} \quad \Gamma.$$

Other boundary conditions can be treated similarly, see, e.g., Girault and Raviart [64, chapter I, §5]. Since the pressure p is only defined up to constant, the condition

(8.1.7) $$\int_\Omega p \, d(x,y) = 0$$

may be imposed to ensure uniqueness. In compact notation the Stokes equations read

(8.1.8) $$\begin{array}{c} -\Delta u + \text{grad } p = f \\ \text{div } u = 0 \end{array} \quad \text{in } \Omega,$$

8.1 The Continuous Problem

with
$$u = 0 \quad \text{on} \quad \Gamma.$$

The Stokes equations give the simplest model for real (and very viscous) fluids. They also appear in the *Navier Stokes equation*, when it is split into a Stokes step and a nonlinear step. An *incompressible flow* is characterized by the condition div $u = 0$, which ensures conservation of mass. The internal forces due to pressure and viscosity (here scaled to be one) balance an external force (modeled by f) such as gravity. We expect pressure of large magnitude at places where otherwise sources or sinks would occur.

Variational Formulation

Functions $u_1, u_2 \in C^2(\Omega) \cap C^0(\overline{\Omega})$ and $p \in C^1(\Omega)$ satisfying (8.1.8) are called *classical solutions*. In general it is relatively difficult to prove the existence of a classical solution. We will see that the story is different for the variational form (weak form) of (8.1.8), Also, this formulation makes it possible to investigate various related problems.

To formulate a weak formulation of (8.1.8), we introduce a velocity test space

$$(8.1.9) \quad X := H_0^1(\Omega) \times H_0^1(\Omega), \quad \text{with} \quad \|v\|_X := \sqrt{\|v_1\|_{H^1}^2 + \|v_2\|_{H^1}^2},$$

and a pressure test space (cf. (8.1.7))

$$(8.1.10) \quad Q := \{q \in L_2(\Omega) : \int_\Omega q \, d(x,y) = 0\}.$$

Then a weak formulation of (8.1.8) reads:

Find $u \in X$ and $p \in Q$ such that

$$(8.1.11) \quad \begin{aligned} a(u,v) + b(v,p) &= \langle f^T v, 1 \rangle_{L_2}, \quad \text{for all } v \in X, \\ b(u,q) &= 0, \quad \text{for all } q \in Q, \end{aligned}$$

where a and b are bilinear forms defined by

$$(8.1.12) \quad \begin{aligned} a(u,v) &:= \int_\Omega (\text{grad}\, u_1)^T \text{grad}\, v_1 + (\text{grad}\, u_2)^T \text{grad}\, v_2 \, d(x,y), \\ b(v,q) &:= -\int_\Omega q \, \text{div}\, v \, d(x,y), \end{aligned}$$

respectively. We like to point out a fundamental difference between the classical and the weak formulation of the Stokes problem. The classical formulation deals with points in the flow domain Ω, whereas the weak formulation deals with functions in appropriate function spaces.

The next lemma collects some auxiliary results. A proof may be found in Hackbusch[76, §12.2].

Lemma 8.1.2 Let $\Omega \subset \mathbb{R}^2$ be a polygonal domain, let $f \in L_2(\Omega) \times L_2(\Omega)$, and let a, b be given by (8.1.12). Then

(a) The bilinear form a is symmetric and continuous on $X \times X$. Moreover, a is X-elliptic, i.e., there exists a constant $\alpha > 0$ such that
$$a(v,v) \geq \alpha \|v\|_X^2, \quad \text{for all } v \in X.$$

(b) The bilinear form b is continuous on $X \times Q$.

(c) The function $F_1(v) := \langle f^T v, 1 \rangle_{L_2}$ is linear and continuous on X, i.e., F_1 is an element of the dual space X' of X.

Part (a) of the above lemma has some immediate consequences. It implies that
$$\|v\|_a := \sqrt{a(v,v)}$$
defines a norm on the Hilbert space X. This norm is known as *energy norm*. It is equivalent to the usual norm $\|\cdot\|_X$. Moreover, we deduce that the bilinear form a is *positive definite*

(8.1.13) $\qquad a(v,v) > 0 \quad \text{for all } v \in X \setminus \{0\}.$

We mention that another common way to derive a weak formulation of the Stokes equation is based upon the idea of restricting the velocity test space X to those functions which satisfy the incompressibility condition div $u = 0$ in Ω

(8.1.14) $\qquad X_0 := \{v \in X : b(v,q) = 0, \text{ for all } q \in Q\}$

(compare Hackbusch [76, §12.2.5]). This time the variational formulation of the Stokes problem reads:

8.1 The Continuous Problem

Find a function $u \in X_0$ such that

(8.1.15) $\qquad a(u,v) = \langle f^T v, 1 \rangle_{L_2}, \quad \text{for all } v \in X_0.$

Observe that in this formulation the pressure p has "disappeared". Moreover, (8.1.15) may be seen as a weak formulation of the Poisson equation

$$-\Delta u = f \quad \text{in } \Omega,$$
$$u = 0 \quad \text{on } \Gamma,$$

with respect to the space X_0 (instead of X). Actually, the mixed formulation (8.1.11) corresponds to the introduction of the Lagrange multiplier p in problem (8.1.15) with respect to X_0.

Also, it is worth noticing that $v \in X_0$ if, and only if there exists a function $\varphi \in H_0^2(\Omega)$ with

$$v = \mathrm{rot}\,\varphi = \left[\frac{\partial \varphi}{\partial y}, -\frac{\partial \varphi}{\partial x}\right]^T.$$

The function φ is the *stream function* connected with the velocity field v. The curves $\{(x,y) : \varphi(x,y) = \text{constant}\}$ are the *streamlines* along which the fluid travels.

Saddle Point Problems

There is a close connection between variational problems of the form (8.1.11) and *saddle point problems*. The theory holds true for quite general Hilbert spaces. Therefore one may interpret, throughout the next two subsections, X and Q as some real Hilbert spaces, a and b as continuous bilinear forms on $X \times X$ and $X \times Q$, respectively. Furthermore, the right hand sides of (8.1.11) may be seen as elements $F_1 \in X'$, $F_2 \in Q'$ of the dual space of X and Q, respectively.

The weak formulation (8.1.11) reads in terms of F_1, F_2:

Find $u \in X$ and $p \in Q$ such that

(8.1.16) $\qquad \begin{aligned} a(u,v) + b(v,p) &= F_1(v), & \text{for all } v \in X, \\ b(u,q) &= F_2(q), & \text{for all } q \in Q. \end{aligned}$

Saddle point problems enter into the picture via the function

(8.1.17) $\qquad J(v,q) := a(v,v) + 2b(v,q) - 2F_1(v) - 2F_2(q).$

The pair (u, p) is called a *saddle point of J* if

(8.1.18) $\qquad J(u, q) \leq J(u, p) \leq J(v, p), \quad \text{for all } v \in X, \ q \in Q.$

The proof of the following theorem may be found in Hackbusch [76, Satz 12.2.4].

Theorem 8.1.3 *Let a be symmetric and X-elliptic. Then $(u, p) \in X \times Q$ is a solution of (8.1.16) if, and only if (u, p) is a saddle point of J defined by (8.1.17).*

Consequently, in view of Lemma 8.1.2, any weak solution of the Stokes problem (8.1.11) is a saddle point of the associated function J.

Existence and Uniqueness of Solutions

To establish an existence result for (8.1.16) or (8.1.18), we need a *compatibility condition* between the spaces X and Q. The condition: there exists a constant $\beta > 0$ such that

(8.1.19) $\qquad \sup_{v \in X \setminus \{0\}} \dfrac{b(v, q)}{\|v\|_X} \geq \beta \|q\|_Q, \quad \text{for all } q \in Q,$

is called *inf-sup condition*. It was introduced independently by Babuška [112] and Brezzi [11].

The proof of the following theorem may be found in the classical paper of Brezzi [11]. It states the existence and uniqueness of a solution for problem (8.1.16) in general Hilbert spaces X and Q.

Theorem 8.1.4 *Let a be a symmetric and positive semidefinite bilinear form on $X \times X$. Then, the saddle point problem (8.1.16) has a unique solution if, and only if a is X_0-elliptic and b fulfills the inf-sup condition (8.1.19).*

8.2 The Discrete Problem

To apply the above theorem to the Stokes case, we have to investigate the inf-sup condition for the associated bilinear form b defined in (8.1.12). Here, the inf-sup condition is related to the shape of Ω. It is satisfied for fairly complicated flow domains. We only state the following lemma (cf. Braess [10, Satz 5.4]).

Lemma 8.1.5 *Let Ω be a polygonal domain. Then the bilinear form b, defined in (8.1.12), satisfies the inf-sup condition (8.1.19).*

This in conjunction with Lemma 8.1.2 and (8.1.13) shows that the Stokes problem has a unique solution.

Corollary 8.1.6 *Let Ω be a polygonal domain. Then, the weak formulation of the Stokes problem (8.1.11) has a unique solution.*

8.2 The Discrete Problem

In this section we will discuss the discretization process of the variational problem (8.1.11) using a finite element method.

Let $Q^h \subset Q$ and $X^h \subset X$ be finite dimensional subspaces of Q and X, respectively. Then the discrete analogue to (8.1.11) reads:

Find $u^h \in X^h$ and $p^h \in Q^h$ such that

(8.2.1)
$$a(u^h, v^h) + b(v^h, p^h) = \langle f^T v^h, 1 \rangle_{L_2}, \quad \text{for all } v^h \in X^h$$
$$b(u^h, q^h) = 0, \quad \text{for all } q^h \in Q^h.$$

This approach is called a mixed (finite element) method. The term mixed refers to the fact that we are looking for independent approximations of both the velocity and the pressure. The index h will eventually refer to a mesh from which these approximations are derived. Furthermore, the system (8.2.1) defines a *conforming approximation* because Q^h and X^h are subspaces of the underlying infinite dimensional spaces.

It should be pointed out that the inclusions $Q^h \subset Q$ and $X^h \subset X$ alone do not guarantee the existence of a stable and meaningful approximation. Again, the crucial assumption is a compatibility condition between Q^h and X^h. We introduce the *discrete inf-sup condition* for the bilinear form b: there exists a constant $\beta_h > 0$ such that

$$(8.2.2) \qquad \sup_{v^h \in X^h \setminus \{0\}} \frac{b(v^h, q^h)}{\|v^h\|_X} \geq \beta_h \|q^h\|_{L_2}, \text{ for all } q^h \in Q^h.$$

Moreover, let

$$X_0^h := \{v^h \in X^h : b(v^h, q^h) = 0, \text{ for all } q^h \in Q^h\}.$$

Then, a is called X_0^h-elliptic if there exists a constant $\alpha_h > 0$ such that

$$(8.2.3) \qquad a(v^h, v^h) \geq \alpha_h \|v^h\|_X, \text{ for all } v^h \in X_0^h.$$

We have the following approximation theorem due to Brezzi [11].

Theorem 8.2.1 *Let (u, p) be the solution of (8.1.11). Furthermore, assume that a is X_0^h-elliptic (cf. 8.2.3) and that b fulfills the discrete inf-sup condition (8.2.2). Then for every h the discrete problem (8.2.1) has a unique solution (u^h, p^h) which satisfies the (optimal) error estimate*

$$\|u - u^h\|_X + \|p - p^h\|_{L_2}$$
$$\leq C_h \left(\inf_{v^h \in X^h} \|u - v^h\|_X + \inf_{q^h \in Q^h} \|p - q^h\|_{L_2} \right).$$

Some remarks are appropriate here. From Lemma 8.1.2(a) we know that for the Stokes problem the bilinear form a is even X-elliptic so that a is X_0^h-elliptic regardless of the choice of the discretization. Moreover, the constant $\alpha = \alpha_h$ is independent of h. On the other hand, however, the discrete inf-sup condition does not follow from the continuous version (8.1.19) and has to be checked in each particular case. The constant C_h does depend on α_h and β_h. It is independent of h if the constants $\alpha = \alpha_h$ and $\beta = \beta_h$ do not depend on h.

8.2 The Discrete Problem

The next step is to introduce basis functions for X^h and Q^h, respectively, and to reformulate (8.2.1) as a matrix problem. Most commonly, both components of the velocity are approximated by the same space. Let $\{N_i\}_{i=1}^n$ denote a "velocity basis" and $\{M_j\}_{j=1}^m$ denote a "pressure basis", i.e.,

$$X^h = \text{span}\{N_1, N_2, \ldots, N_n\} \times \text{span}\{N_1, N_2, \ldots, N_n\}$$

(8.2.4) $$= \text{span}\left\{ \begin{bmatrix} N_1 \\ 0 \end{bmatrix}, \begin{bmatrix} N_2 \\ 0 \end{bmatrix}, \ldots, \begin{bmatrix} N_n \\ 0 \end{bmatrix}, \begin{bmatrix} 0 \\ N_1 \end{bmatrix}, \ldots, \begin{bmatrix} 0 \\ N_n \end{bmatrix} \right\},$$

$$Q^h = \text{span}\{M_1, M_2, \ldots, M_m\}.$$

In other words, the functions $u^h = [u_1^h\ u_2^h]^T \in X^h$ and $p^h \in Q^h$

(8.2.5)
$$u_1^h(x,y) = \sum_{i=1}^n \alpha_i N_i(x,y), \quad u_2^h(x,y) = \sum_{i=1}^n \beta_i N_i(x,y),$$
$$p^h(x,y) = \sum_{j=1}^m \gamma_j M_j(x,y),$$

are uniquely defined by the vectors

(8.2.6) $\quad u_1^h := [\alpha_1 \ \cdots \ \alpha_n]^T, \ u_2^h := [\beta_1 \ \cdots \ \beta_n]^T, \ p^h := [\gamma_1 \ \cdots \ \gamma_m]^T.$

For convenience we use the same notation for the function and for the coefficient vector in its basis representation. The actual meaning of u^h and p^h will be clear from the context.

Since a and b are linear in both components, it suffices to enforce (8.2.1) for the basis functions of X^h and Q^h, respectively. We arrive at the following equivalent formulation:

Find vectors $u_1^h, u_2^h \in \mathbb{R}^n$, and $p^h \in \mathbb{R}^m$ with

(8.2.7)
$$a(u^h, [N_i\ 0]^T) + b([N_i\ 0]^T, p^h) = \langle [N_i\ 0]f, 1\rangle_{L_2}$$
$$a(u^h, [0\ N_i]^T) + b([0\ N_i]^T, p^h) = \langle [0\ N_i]f, 1\rangle_{L_2}$$
$$b(u^h, M_j) = 0,$$

for $i = 1, 2, \ldots, n$, and $j = 1, 2, \ldots, m$.

Next, let us define matrices

$$A_1 := [a_1^{i,j}] = [a_2^{i,j}] =: A_2 \in \mathbb{R}^{n\times n},$$
$$B_1 := [b_1^{j,i}] \in \mathbb{R}^{m\times n}, \quad B_2 := [b_2^{j,i}] \in \mathbb{R}^{m\times n},$$

and vectors

$$f_1^h := (f_1^i) \in \mathbb{R}^n, \quad f_2^h := (f_2^i) \in \mathbb{R}^n,$$

where

$$a_1^{i,j} = a_2^{i,j} = \int_\Omega \frac{\partial N_i}{\partial x}\frac{\partial N_j}{\partial x} + \frac{\partial N_i}{\partial y}\frac{\partial N_j}{\partial y}\, d(x,y),$$

$$b_1^{i,j} = -\int_\Omega M_j \frac{\partial N_i}{\partial x}\, d(x,y), \quad b_2^{i,j} = -\int_\Omega M_j \frac{\partial N_i}{\partial y}\, d(x,y).$$

and

$$f_1^i = \int_\Omega f_1 N_i\, d(x,y), \quad f_2^i = \int_\Omega f_2 N_i\, d(x,y).$$

Now, we are in a position to rewrite (8.2.1) (or (8.2.7)) as a matrix problem: Find vectors $u_1^h, u_2^h \in \mathbb{R}^n$, and $p^h \in \mathbb{R}^m$ with

(8.2.8)
$$\begin{bmatrix} A_1 & 0 & B_1^T \\ 0 & A_2 & B_2^T \\ B_1 & B_2 & 0 \end{bmatrix} \begin{bmatrix} u_1^h \\ u_2^h \\ p^h \end{bmatrix} = \begin{bmatrix} f_1^h \\ f_2^h \\ 0 \end{bmatrix}.$$

Alternatively, and more commonly, this system may be written in the form

(8.2.9)
$$\begin{bmatrix} A & B^T \\ B & 0 \end{bmatrix} \begin{bmatrix} u^h \\ p^h \end{bmatrix} = \begin{bmatrix} f^h \\ 0 \end{bmatrix}.$$

The Linear System

Theorem 8.2.1 states that, if a is X_0^h-elliptic (cf. 8.2.3) and if b fulfills the discrete inf-sup condition (8.2.2), then the discrete problem (8.2.1) has a unique solution. This, of course, implies that the linear system (8.2.8) has a unique solution. In this section we will interpret the ellipticity condition and the inf-sup condition in terms of matrices and show why these conditions eventually lead to a nonsingular matrix (8.2.8).

8.2 The Discrete Problem

To begin with, let us investigate the matrix

$$A = \begin{bmatrix} A_1 & 0 \\ 0 & A_2 \end{bmatrix},$$

which is known as *stiffness matrix*. The ellipticity (cf. Lemma 8.1.2(a)) of the bilinear form a implies that the stiffness matrix is positive definite. To see this, recall that

$$a_1^{i,j} = a_2^{i,j} = a([N_i \; 0]^T, [N_j \; 0]^T) = a([0 \; N_i]^T, [0 \; N_j]^T),$$

and observe that with (8.2.5) and (8.2.6)

(8.2.10) $\quad [(u_1^h)^T \; (u_2^h)^T] A \begin{bmatrix} u_1^h \\ u_2^h \end{bmatrix} = a(u^h, u^h) \geq \alpha \|u^h\|_X > 0.$

Moreover, the matrix A may be seen as the discrete representation of a second-order operator. The Dirichlet boundary conditions (8.1.6) ensure that the smallest eigenvalue of A is bounded from below by a constant C times the square of the grid size

(8.2.11) $\quad \min_{u^h \neq 0} \dfrac{(u^h)^T A u^h}{(u^h)^T u^h} \geq Ch^2$

(cf. Axelsson and Barker [8, p. 240]).

Since A is positive definite, the coefficient matrix (8.2.9) is nonsingular if, and only if B has full rank. Here, the discrete inf-sup condition (8.2.2) enters into the picture. Let us rewrite this condition in terms of matrices (compare Brezzi and Fortin [12, pp. 75]). It is straightforward to verify that

(8.2.12) $\quad \|q^h\|_{L_2}^2 = (q^h)^T \mathcal{Q}_p q^h,$

where

(8.2.13) $\quad \mathcal{Q}_p := [\langle M_i, M_j \rangle_{L_2}]_{i,j=1,2,\ldots,m}$

denotes the *pressure mass matrix*. It is a Grammian matrix of basis functions for Q^h and hence it is symmetric and positive definite. Actually, the condition number of $\mathcal{Q}_p$ is independent of h for any usual finite element basis (cf. Wathen [128]). Furthermore, we have in view of (8.1.4) and (8.1.12)

(8.2.14) $\quad \|v^h\|_X^2 \geq a(v^h, v^h).$

and

(8.2.15) $$b(v^h, q^h) = (q^h)^T B v^h.$$

Following Wathen and Silvester [130] we will now derive a lower bound for the smallest singular value $\sigma_1(B)$ of B. Starting from (8.2.14) and (8.2.11) we obtain

$$\|v^h\|_X^2 \geq a(v^h, v^h) = (v^h)^T A v^h \geq C h^2 (v^h)^T v^h,$$

which together with the discrete inf-sup condition, (8.2.12), and (8.2.15) implies

$$\sigma_1(B) = \min_{q^h \in \mathbb{R}^m \setminus \{0\}} \max_{v^h \in \mathbb{R}^{2n} \setminus \{0\}} \frac{(q^h)^T B v^h}{\sqrt{(q^h)^T q^h} \sqrt{(v^h)^T v^h}}$$

$$\geq h \beta_h \sqrt{C} \min_{q^h \in \mathbb{R}^m \setminus \{0\}} \frac{\sqrt{(q^h)^T Q_p q^h}}{\sqrt{(q^h)^T q^h}}.$$

In other words, B has full rank. We arrive at the following variant of Theorem 8.2.1.

Theorem 8.2.2 *If a is X_0^h-elliptic (cf. 8.2.3) and if b fulfills the discrete inf-sup condition (cf. 8.2.2), then the "discrete" Stokes system (8.2.9) has a unique solution.*

In fact, the special structure of the coefficient matrix in the Stokes system (8.2.9) may be viewed as a template for symmetric and (highly) indefinite linear systems, as is apparent from the next theorem.

Theorem 8.2.3 *Let $A \in \mathbb{R}^{l \times l}$ and $B \in \mathbb{R}^{m \times l}$ with $l \geq m$. If A is symmetric and positive definite and if B has full rank, then*

$$\mathcal{A} = \begin{bmatrix} A & B^T \\ B & 0 \end{bmatrix}$$

has l positive and m negative eigenvalues.

Proof. We apply block Gaussian elimination to $\mathcal{A}$ to obtain the following block LDL^T decomposition

$$\mathcal{A} = \begin{bmatrix} I_l & 0 \\ BA^{-1} & I_m \end{bmatrix} \begin{bmatrix} A & 0 \\ 0 & -BA^{-1}B^T \end{bmatrix} \begin{bmatrix} I_l & 0 \\ BA^{-1} & I_m \end{bmatrix}^T.$$

Since A^{-1} is positive definite and B has full rank the so-called *Schur-complement* $BA^{-1}B^T \in R^{m \times m}$ is positive definite

$$x^T B A^{-1} B^T x = (B^T x)^T A^{-1} B^T x = y^T A^{-1} y > 0, \quad x \in \mathbb{R}^m \setminus \{0\}.$$

Here, we used the fact that $B^T x = 0$ is only possible for $x = 0$. The statement follows from Sylvesters law of inertia (see, e.g., Golub and Van Loan [71, Theorem 8.1.12]). □

8.3 Some Finite Element Spaces

In this section we will introduce and discuss some commonly used finite element spaces. For a complete list, see for example Girault and Raviart [64, chapter 2] and Gunzburger [74, chapter 3].

Let us start with a formal definition of a *finite element*. It is a triple $(T, \mathcal{P}_T, \Sigma)$, where

T is a geometric object, e.g., a triangle,

$\mathcal{P}_T$ is a finite dimensional linear space of functions defined on T,

Σ is "set of degrees of freedom", such that a function $p \in \mathcal{P}_T$ is uniquely defined by the degrees of freedom Σ, e.g., the values at the vertices of T.

Most commonly, the finite element spaces consist of piecewise polynomial functions on subdivisions or *triangulations* of the polygon Ω into elements T_j. These elements will be triangles, rectangles or quadrilaterals. We will discuss in greater detail the case of triangles. The analysis for rectangular elements is entirely analogous.

A triangulation $\mathcal{T}_h = \{T_1, T_2, \ldots, T_k\}$ of Ω is called *admissible* if

$$\overline{\Omega} = \bigcup_{i=1}^{k} T_i$$

and any two elements are either disjoint or share exactly either one side or one vertex.

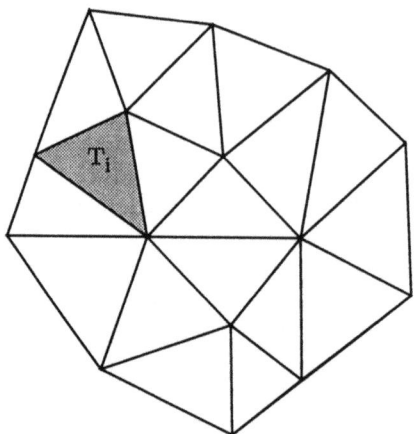

Figure 8.3.1 Admissible triangulation.

The parameter h is a measure for the size of the elements in $\mathcal{T}_h$. More precisely, let h_i denote the diameter of T_i, then

$$h := \max_{i=1,2,\ldots,k} h_i.$$

Furthermore, let ρ_i denote the diameter of the largest inscribed disk in T_i. Then, a family $\{\mathcal{T}_h\}_h$ of admissible triangulations of $\overline{\Omega}$ is said to be *regular* if there exists a constant σ with

$$\frac{h_i}{\rho_i} \leq \sigma \quad \text{for all} \quad T_i \in \mathcal{T}_h \quad \text{and all} \quad h.$$

As usual, p is called a piecewise polynomial on $\mathcal{T}_h$, if the restrictions $p|_{T_i}$ on the elements of $\mathcal{T}_h$ are polynomials. The next lemma is a useful guide for the design of conforming methods. A proof may be found in Braess [10, Satz 5.2].

Lemma 8.3.2 *Let $\mathcal{T}_h$ be an admissible triangulation of Ω and let p be a piecewise polynomial on $\mathcal{T}_h$. Then, $p \in H^1(\Omega)$ if, and only if p is a continuous function on $\overline{\Omega}$.*

8.3 Some Finite Element Spaces

To discuss conditions under which a piecewise polynomial is continuous on $\overline{\Omega}$, consider the polynomial function space

(8.3.1) $\qquad \mathcal{P}_n := \{p_n(x,y) = \sum_{i+j \leq n,\ i,j \geq 0} c_{ij} x^i y^j,\ c_{ij} \in \mathbb{R}\}.$

Note that dim $\mathcal{P}_n = (n+1)(n+2)/2$. Here, we are only interested in the cases $n = 1, 2$, i.e., we will investigate piecewise linear

$$\mathcal{P}_1 = \{p_1(x,y) = c_{00} + c_{10}x + c_{01}y,\ c_{ij} \in \mathbb{R}\},$$

and piecewise quadratic polynomials

$$\mathcal{P}_2 = \{p_2(x,y) = c_{00} + c_{10}x + c_{20}x^2 + c_{11}xy + c_{01}y + c_{02}y^2,\ c_{ij} \in \mathbb{R}\},$$

respectively.

We start with the case $n = 1$. We are looking for a suitable representation of a continuous and piecewise linear polynomial p_1 on the admissible triangulation $\mathcal{T}_h = \{T_1, T_2, \ldots, T_k\}$. That is, $p_1 \in C^0(\overline{\Omega})$ and

$$p_1(x,y) = p_1^{(l)}(x,y) \quad \text{for} \quad (x,y) \in T_l, \quad p_1^{(l)} \in \mathcal{P}_1, \quad l = 1, 2, \ldots, k.$$

Let $(x_j^{(l)}, y_j^{(l)})$, $j = 1, 2, 3$, denote the vertices (or *node points*) of a triangle T_l (compare Figure 8.3.3(a)).

Figure 8.3.3 Interpolation points for (a) $p_1^{(l)}$ and (b) $p_2^{(l)}$.

Obviously, the coefficients of $p_1^{(l)}(x,y) = c_{00} + c_{10}x + c_{01}y \in \mathcal{P}_1$ are uniquely determined by the values at the vertices $p_1^{(l)}(x_j^{(l)}, y_j^{(l)})$, $j = 1, 2, 3$. It is convenient to introduce the so-called *nodal basis functions* with respect to the element T_l. These are linear polynomials $N_j^{(l)} \in \mathcal{P}_1$ defined by the interpolatory conditions

$$N_j^{(l)}(x_i^{(l)}, y_i^{(l)}) = \begin{cases} 1 & \text{for } i = j \\ 0 & \text{for } i \neq j \end{cases}, \quad i, j = 1, 2, 3.$$

They may be viewed as fundamental polynomials of interpolation at the points $(x_i^{(l)}, y_i^{(l)})$. We have the following representation

(8.3.2) $$p_1^{(l)}(x, y) = \sum_{j=1}^{3} p_1^{(l)}(x_j^{(l)}, y_j^{(l)}) N_j^{(l)}(x, y).$$

Actually, since $p_1^{(l)}$ is a linear function on each side of the triangle, its function values on a side are already determined by its function values at the corresponding vertices. In the light of this observation, it is straightforward to write down a basis representation for p_1. Let (x_j, y_j), $j = 1, 2, \ldots, m$, denote the vertices of $\mathcal{T}_h$ and let

$$N_j(x_i, y_i) = \begin{cases} 1 & \text{for } i = j \\ 0 & \text{for } i \neq j \end{cases}, \quad i, j = 1, 2, \ldots, m.$$

denote the nodal basis functions for $\mathcal{T}_h$. The global nodal function N_j is just the collection of the local nodal functions $N_j^{(l)}$ on those triangles which have the node (x_j, y_j) in common.

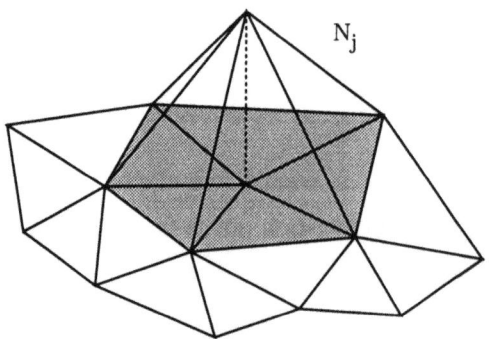

Figure 8.3.4 The basis function N_j.

8.3 Some Finite Element Spaces

Then it follows directly from (8.3.2) that

$$p_1(x,y) = \sum_{j=1}^m p_1(x_j, y_j) N_j(x,y).$$

Observe that the dimension of the space of continuous and piecewise linear polynomials on $\mathcal{T}_h$ is given by the number of vertices in $\mathcal{T}_h$.

Likewise, it is not hard to prove that a quadratic polynomial $p_2^{(l)} \in \mathcal{P}_2$ is uniquely defined by its function values on the vertices $(x_j^{(l)}, y_j^{(l)})$, $j = 1, 2, 3$, and on the midpoints $(x_{12}^{(l)}, y_{12}^{(l)}), (x_{13}^{(l)}, y_{13}^{(l)}), (x_{23}^{(l)}, y_{23}^{(l)})$ of the sides of a triangle T_l (compare Figure 8.3.3(b))

$$\begin{aligned}
p_2^{(l)}(x,y) = &\sum_{j=1}^3 p^{(l)}(x_j^{(l)}, y_j^{(l)}) N_j^{(l)}(x,y)(2N_j^{(l)}(x,y) - 1) \\
&+ \sum_{i,j=1,\ i<j}^3 p^{(l)}(x_{ij}^{(l)}, y_{ij}^{(l)}) 4 N_i^{(l)}(x,y) N_j^{(l)}(x,y).
\end{aligned}$$
(8.3.3)

Here, we used the fact that $N_j^{(l)}(x_{ij}^{(l)}, y_{ij}^{(l)}) = N_i^{(l)}(x_{ij}^{(l)}, y_{ij}^{(l)}) = 1/2$. The representation (8.3.3) implicitly defines nodal basis functions from $\mathcal{P}_2$ with respect to T_l. Working from here, one may derive a representation for a continuous and piecewise quadratic polynomial p_2 on $\mathcal{T}_h$ in terms of global nodal basis functions. This time, however, the dimension of the associated space is given by the number of vertices plus the number of midpoints of elements in $\mathcal{T}_h$.

The next lemma summarizes some properties of piecewise continuous functions. For convenience we introduce the space

$$V^j(\mathcal{T}_h) := \{p_j \in C^0(\overline{\Omega}) : p_j|_{T_l} \in \mathcal{P}_j, l = 1, 2, \ldots, k\}$$

and

$$V_0^j(\mathcal{T}_h) := \{p_j \in C^0(\overline{\Omega}) : p_j = 0 \text{ on } \Gamma, p_j|_{T_l} \in \mathcal{P}_j, l = 1, 2, \ldots, k\},$$

respectively.

Lemma 8.3.5 Let $T_h = \{T_1, T_2, \ldots, T_k\}$ be an admissible triangulation of Ω. Then

(a) $V^j(T_h) \subset H^1(\Omega)$ and $V_0^j(T_h) \subset H_0^1(\Omega)$.

(b) Each function in $V^1(T_h)$ is uniquely determined by its function values at the node points of T_h. Each function in $V_0^1(T_h)$ is uniquely determined by its function values at the node points of T_h which do not belong to the boundary Ω.

(c) Each function in $V^2(T_h)$ is uniquely determined by its function values at the node points of T_h and of the midpoints of all sides of the triangles in T_h. Each function in $V_0^2(T_h)$ is uniquely determined by its function values at those node points of T_h and of those midpoints of all sides of the triangles in T_h which do not belong to the boundary Ω.

Now we are in a position to introduce some mixed finite elements. According to the previous section we have to define basis functions for Q^h and X^h (cf. 8.2.4), respectively.

We pick two popular elements and discuss some of the issues in connection with choosing a mixed finite element method.

We start by introducing the so-called *Hood-Taylor method* (cf. Girault and Raviart [64, §4.2]). Here, the approximation spaces are based on the same grid but on different degree polynomials

(8.3.4)
$$X^h := V_0^2(T_h) \times V_0^2(T_h),$$
$$Q^h := V^1(T_h) \cap Q.$$

A method, closely related to the Hood-Taylor method, is known under the name $P_1 isoP_2$ method (cf. Brezzi and Fortin [12, pp. 254]). Here the piecewise linear pressure element is coupled with a piecewise linear velocity element. However, the triangulation for the velocity $T_{h/2}$ is obtained from T_h by subdividing each triangle T_l into four triangles by joining the midsides. Then define

(8.3.5)
$$X^h := V_0^1(T_{h/2}) \times V_0^1(T_{h/2}),$$
$$Q^h := V^1(T_h) \cap Q.$$

8.3 Some Finite Element Spaces

We remark that in practice the constraint $\int_\Omega p_1 \, d(x,y) = 0$ is ignored. It is imposed a posteriori by subtracting, from the computed pressure, its mean.

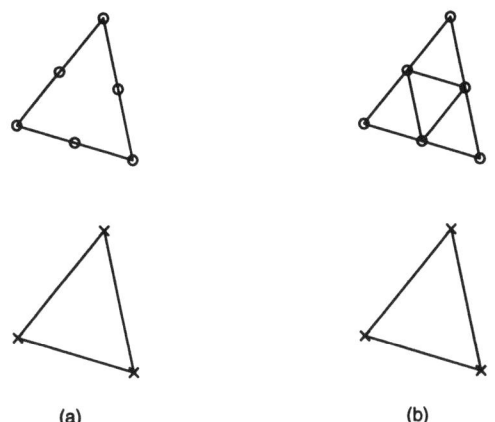

Figure 8.3.6 (a) Hood-Taylor Element, (b) $P_1 isoP_2$ element. (○) degrees of freedom for the velocity; (×) degrees of freedom for the pressure.

Next, we briefly discuss the case of rectangular elements. It is completely analogous to the triangular case. Only, we have to substitute the space $\mathcal{P}_n$ by
$$\mathcal{D}_n := \{p_n(x,y) = \sum_{0 \leq i,j \leq n} c_{ij} x^i y^j, \; c_{i,j} \in \mathbb{R}\},$$
Note that $\dim \mathcal{D}_n = (n+1)^2$. This time the coefficients of
$$p_1(x,y) = c_{00} + c_{10}x + c_{01}y + c_{11}xy \in \mathcal{D}_1$$
are uniquely defined by its function values at the vertices of a quadrilateral. To determine a quadratic polynomial $p_2 \in \mathcal{D}_2$ we need 9 degrees of freedom
$$p_2(x,y) = c_{00} + c_{10}x + c_{01}y + c_{11}xy + c_{20}x^2 + c_{02}y^2 + c_{21}x^2y + c_{12}xy^2 + c_{22}x^2y^2.$$

For the *Hood-Taylor element* these are the 4 vertices, the 4 midpoint of the sides, and the midpoint of the rectangle. To obtain the velocity approximation for the method $Q_1 isoQ_2$ element each rectangles is subdivided into four rectangles by joining the midsides.

Both methods lead to stable schemes (cf. Bercovier and Pironneau [9]).

Theorem 8.3.7 Let (u,p) be the solution of the Stokes problem (8.1.11). Furthermore, let T_h be a regular triangulation of Ω, which is either based on triangles or rectangles.

(a) Let X^h and Q^h be chosen as in (8.3.4) or (8.3.5), respectively. Then, the bilinear form b (cf. (8.2.1)) fulfills the discrete inf-sup condition

$$\sup_{v^h \in X^h \setminus \{0\}} \frac{b(v^h, q^h)}{\|v^h\|_X} \geq \beta \|q^h\|_{L_2}, \text{ for all } q^h \in Q^h,$$

where the constant β is independent of h (cf. Theorem 8.2.1).

(b) Let X^h and Q^h be chosen as in (8.3.4) and let (u^h, p^h) be the solution of the discrete problem (8.2.1), then

$$\|u - u^h\|_X + \|p - p^h\|_{L_2} \leq \mathcal{O}(h^2).$$

(b) Let X^h and Q^h be chosen as in (8.3.5) and let (u^h, p^h) be the solution of the discrete problem (8.2.1), then

$$\|u - u^h\|_X + \|p - p^h\|_{L_2} \leq \mathcal{O}(h).$$

Example 8.3.8 Let us now demonstrate the performance of MINRES when applied to a typical Stokes system. As a test example we solved the "leaky" two-dimensional lid-driven cavity problem in a unit square domain with point body forces (represented by a random vector f for points in the domain interior) and no flow boundary conditions (cf. Peyret and Taylor [102]). As a representative mixed approximation we used the $Q1isoQ2$ elements based on squares. †

Let e denote the number of elements along each side for the pressure approximation, i.e., the velocity approximation is based on $2e$ elements along each

---

† The author would like to thank David Silvester for generating the matrices.

8.3 Some Finite Element Spaces

side. Then the dimension of the matrices $A_e \in \mathbb{R}^{2n \times 2n}$ and $B_e \in \mathbb{R}^{m \times 2n}$ in

$$\mathcal{A}_e = \begin{bmatrix} A_e & B_e^T \\ B_e & 0 \end{bmatrix}$$

is given by

$$2n = 2(2e+1)^2 \quad \text{and} \quad m = (e+1)^2,$$

if we ignore the no flow boundary condition. In fact, the boundary conditions were imposed, after having assembled the matrices A_e and B_e, by scaling the corresponding elements of A_e and of the righthand side, respectively. This approach is convenient in terms of programming work. It, however, leads to larger systems.

We did not apply MINRES directly to the matrix $\mathcal{A}_e$, but to the preconditoned matrix $\mathcal{M}_e^{-1} \mathcal{A}_e$, where the (diagonal) preconditioner

$$\mathcal{M}_e = \begin{bmatrix} M_A & 0 \\ 0 & M_\mathcal{Q} \end{bmatrix}$$

is defined by the diagonal M_A of A_e and the diagonal $M_\mathcal{Q}$ of the associated pressure mass matrix $\mathcal{Q}_p$ (cf. 8.2.13). This type of preconditioner was investigated in great detail by Wathen and Silvester [130].

The next table displays the mesh size parameter $h = 1/2e$, the number N of unknowns, computed bounds of the eigenvalues of the preconditioned system $\mathcal{M}_e^{-1} \mathcal{A}_e$

$$\lambda_{-m} \leq \cdots \leq \lambda_{-1} < 0 < \lambda_1 \leq \cdots \leq \lambda_{2n},$$

and the resulting asymptotic rate of convergence κ which was computed with asympfac.

h	N	λ_{-m}	λ_{-1}	λ_1	λ_{2n}	κ
1/8	187	-0.6998	-0.0787	0.1954	1.8271	0.8967
1/16	659	-0.7277	-0.0737	0.0557	1.8555	0.9462
1/32	2467	-0.7351	-0.0399	0.0141	1.8641	0.9799

Table 8.3.9 Characteristic numbers for scaled Stokes matrices $\mathcal{M}_e^{-1} \mathcal{A}_e$.

We would like to mention that Wathen and Silvester [130] proved that the spectrum of $\mathcal{M}_e^{-1}\mathcal{A}_e$ is contained in

$$[-a, -bh] \cup [ch^2, d],$$

where $a, b, c,$ and d are positive constants (independent of h). Based on this inclusion sets and the analysis outlined in Section 3.3, Wathen, Fischer and Silvester [129] recently showed that the asymptotic convergence factor κ (3.1.5) may be bounded as follows

$$\kappa\left([-a, -bh] \cup [ch^2, d]\right) \leq 1 - \sqrt{bc/ad}\ h^{3/2} + \mathcal{O}(h^{5/2}),$$

if h is small enough. This compares with an asymptotic convergence rate

$$\kappa\left([ch^2, d]\right) \leq 1 - 2\sqrt{c/d}\ h^1 + \mathcal{O}(h^2),$$

for MINRES applied to a positive definite system and to

$$\kappa\left([c^2h^4, d^2]\right) \leq 1 - 2(c/d)\ h^2 + \mathcal{O}(h^4),$$

for the "normal equation case", respectively.

Let us now report on some numerical results. The vector $x_0 = 0$ was selected as initial approximation. In each case, the iteration was considered to converge when the nth residual r_n^{MR} satisfied

$$\frac{\|r_n^{\text{MR}}\|_2}{\|r_0\|_2} \leq 10^{-10}.$$

For a discussion of these results and for a comparison of the performance of various alternative iterative strategies for the solution of the Stokes problem we refer to Wathen, Fischer and Silvester [129] and Elman [36], respectively.

8.3 Some Finite Element Spaces

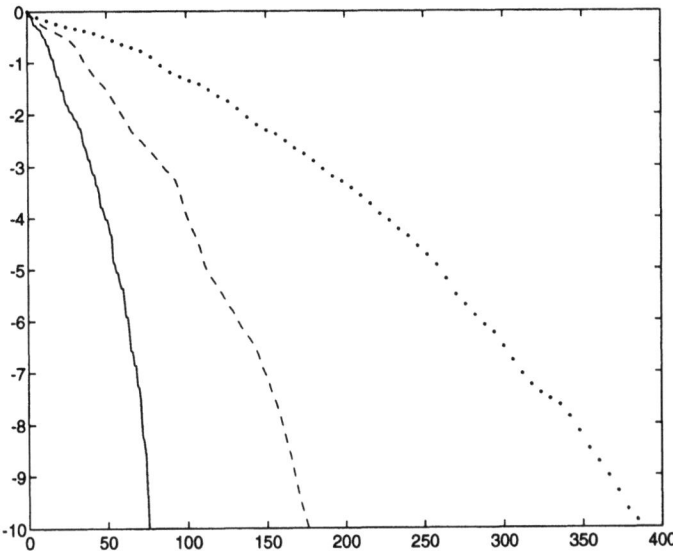

Figure 8.3.10 Residual norms $\log_{10} \|f - Ax_n^{\mathrm{MR}}\|_2/\|r_0\|_2$ versus the number of iterations n for the matrix $\mathcal{M}_e^{-1}\mathcal{A}_e$. Solid curve: $h = 1/8$; dashed curve: $h = 1/16$; dotted curve: $h = 1/32$.

□

9 Approximating the A-Norm

In this chapter we devise a method for the approximation of the A-norm of the error $\|x_* - x_n\|_A$ without actually computing the solution x_*.

9.1 Energy Norm

The CG process (cf. Section 6.2) is often introduced as a minimization scheme. Here, we will briefly outline this approach. Furthermore, we will indicate that the A-norm of the current iterate represents the "energy" of a certain functional.

The quadratic form

$$(9.1.1) \qquad g(x) := \frac{1}{2} x^T A x - x^T f$$

has precisely one stationary point

$$\operatorname{grad} g(x) = Ax - f = 0,$$

namely the solution x_* of $Ax = f$. In other words, if A is symmetric and positive definite, then x_* is the point where g attains its minimum. Thus, instead of solving the linear system $Ax = f$ one may as well minimize the functional (9.1.1). Observe, that

$$g(x) = \frac{1}{2}(x - x_*)^T A (x - x_*) - \frac{1}{2} x_*^T A x_*.$$

Since the last term is constant, minimizing g is equivalent to minimize the error functional

$$E(x) := (x - x_*)^T A(x - x_*) = \|x - x_*\|_A^2.$$

This is a familiar expression. In Example 2.4.8 we showed that the CG method minimizes E with respect to an associate Krylov subspace, where $E(x_n^{CG}) = \|\varepsilon_n^{CG}\|_A^2$.

9.1 Energy Norm

Let us briefly indicate that g may be viewed as the discrete version of an *energy functional*. Based on this connection, the A-norm of a vector is sometimes referred to as *energy norm*. To this end consider Poisson's equation:

Find a function $u : \Omega \to \mathbb{R}$ such that

(9.1.2)
$$-\Delta u = f, \quad \text{in } \Omega,$$
$$u = 0, \quad \text{on } \Gamma,$$

with homogeneous boundary conditions. Here, Γ denotes the boundary of an open, bounded, and connected region $\Omega \subset \mathbb{R}^2$. The partial differential equation (9.1.2) may be seen as a (simplified) description of a physical problem. An elastic membrane is fixed round a contour Γ. It is subject to some load (vertical force) f. The problem is to find the deflection (vertical displacement) u of the membrane throughout Ω.

It is again helpful to reformulate the elliptic PDE (9.1.2) as a variational problem (compare (8.1.15)):

Find a function $u \in H_0^1(\Omega)$ such that

(9.1.3) $\qquad a(u, v) = \langle f, v \rangle_{L_2}, \quad \text{for all} \quad v \in H_0^1(\Omega),$

where (compare (8.1.12))

$$a(u, v) = \int_\Omega (\operatorname{grad} u)^T \operatorname{grad} v \, d(x, y).$$

This time the saddle point problem (cf. (8.1.17)) turns into a minimization problem. With

$$J(v) := \frac{1}{2} a(v, v) - \langle f, v \rangle_{L_2}$$

it is a standard piece of analysis to show (cf. Hackbusch [76, Satz 6.5.12]) that u solves (9.1.3) if, and only if

$$J(u) \leq J(v), \quad \text{for all} \quad v \in H_0^1(\Omega).$$

Here $J(v)$ represents the *total potential energy* associated with the displacement v.

We have already shown (cf. (8.2.10)) that

$$J(v^h) = \frac{1}{2}x^T A x - x^T f^h,$$

where $v^h(x,y) = \sum_{i=1}^{N} x_i N_i(x,y)$ is an element from a finite dimensional subspace of $H_0^1(\Omega)$, $f_i^h = \langle f, N_i \rangle_{L_2}$, and A is the associated positive definite stiffness matrix.

Therefore $g(x) = J(v^h)$ may be seen as "discrete version" of $J(v)$ and $\|x - x_*\|_A^2$ is essentially the potential energy of v^h.

Hence, from a physical point of view it is interesting to measure the A-norm of the current approximation. How this task can be accomplished, without actually computing the solution x_*, is subject of the next sections.

9.2 Approximating the A-Norm of the Error

We divide the discussion in two parts. First we treat the CG iteration, which has some special properties. Second, we comment on general polynomial iteration methods.

CG Case

Compared to other polynomial iteration methods, the CG method (cf. Section 6.2)

$$\varepsilon_n^{CG} = \psi_n^{OR}(A)\varepsilon_0$$

has the distinguished feature of minimizing the A-norm of the error with respect to a certain Krylov subspace. Recall, that the residual polynomials are uniquely defined by the orthogonality relation

(9.2.1) $$\langle \psi_k^{OR}, \psi_l^{OR} \rangle_{GAL} = r_0^T \psi_k^{OR}(A) \psi_l^{OR}(A) r_0 = 0, \quad k \neq l,$$

and by the interpolatory condition

(9.2.2) $$\psi_n^{OR}(0) = 1.$$

The next lemma collects some special properties of CG iterates.

9.2 Approximating the A-Norm of the Error

Lemma 9.2.1 *Let x_n^{CG} denote a CG iterate and let $\varepsilon_n^{CG} = x_* - x_n^{CG}$ denote the corresponding error with respect to the linear system $Ax_* = f$. Then, we have*

(a) $\qquad (\varepsilon_k^{CG})^T A(\varepsilon_j^{CG} - \varepsilon_i^{CG}) = 0, \quad \text{for} \quad k \geq j, i.$

(b) $\qquad (x_{k+1}^{CG} - x_k^{CG})^T A(x_{j+1}^{CG} - x_j^{CG}) = 0, \quad \text{for} \quad k > j.$

Proof. (a) In view of (9.2.2) we have that $(\psi_j^{OR}(t) - \psi_i^{OR}(t))/t$ is a polynomial of degree at most $k-1$. Hence, we deduce from (9.2.1)

$$(\varepsilon_k^{CG})^T A(\varepsilon_j^{CG} - \varepsilon_i^{CG}) = \langle \psi_k^{OR}(t), (\psi_j^{OR}(t) - \psi_i^{OR}(t))/t \rangle_{\text{GAL}} = 0.$$

The proof of (b) follows directly from (a)

$$(x_{k+1}^{CG} - x_k^{CG})^T A(x_{j+1}^{CG} - x_j^{CG}) = (\varepsilon_k^{CG} - \varepsilon_{k+1}^{CG})^T A(\varepsilon_j^{CG} - \varepsilon_{j+1}^{CG}) = 0.$$

$\square$

The following theorem provides computable approximations for the A-norm of the error and is due to Deuflhard [32].

Theorem 9.2.2 *Let $\varepsilon_n^{CG} = x_* - x_n^{CG}$ denote the error of the nth CG iterate x_n^{CG} with respect to the linear system $Ax_* = f$. Moreover, let η_{k+1}^{CG} denote the steplength (6.2.7) in the CG method. Then, we have*

(a) $\qquad \|\varepsilon_n^{CG}\|_A^2 = \|\varepsilon_0^{CG}\|_A^2 - \|x_n^{CG} - x_0^{CG}\|_A^2.$

(b) $\qquad \|x_n^{CG} - x_0^{CG}\|_A^2 = \sum_{k=0}^{n-1} \|x_{k+1}^{CG} - x_k^{CG}\|_A^2.$

(c) $\qquad \|x_{k+1}^{CG} - x_k^{CG}\|_A^2 = |\eta_{k+1}^{CG}|(r_k^{CG})^T r_k^{CG}.$

Proof. (a) We apply Lemma 9.2.1(a) to obtain

$$\|\varepsilon_n^{CG}\|_A^2 - \|\varepsilon_0^{CG}\|_A^2 + \|x_n^{CG} - x_0^{CG}\|_A^2 = 2(\varepsilon_n^{CG})^T A \varepsilon_n^{CG} - 2(\varepsilon_0^{CG})^T A \varepsilon_n^{CG} = 0.$$

The proof of (b) is based upon Lemma 9.2.1(b)

$$\|x_n^{CG} - x_0^{CG}\|_A^2 = \|\sum_{k=0}^{n-1}(x_{k+1}^{CG} - x_k^{CG})\|_A^2 = \sum_{k=0}^{n-1} \|x_{k+1}^{CG} - x_k^{CG}\|_A^2.$$

Finally, part (c) follows from (6.2.8)

$$\|x_{k+1}^{CG} - x_k^{CG}\|_A^2 = (\eta_{k+1}^{CG})^2 \|w_k\|_A^2 = |\eta_{k+1}^{CG}|(r_k^{CG})^T r_k^{CG}.$$

□

Some remarks are appropriate here. The computation of $\|x_n^{CG} - x_0^{CG}\|_A$ requires the knowledge of η_{k+1}^{CG} and of the inner product $(r_k^{CG})^T r_k^{CG}$. Recall that in course of the CG iteration (cf. Section 6.2) these quantities have to be computed anyway. In other words, the evaluation of $\|x_n^{CG} - x_0^{CG}\|_A$ (essentially) comes for free. Also, we have the following inequalities

$$\|x_n^{CG} - x_0^{CG}\|_A < \|x_{n+1}^{CG} - x_0^{CG}\|_A < \cdots < \|x_L^{CG} - x_0^{CG}\|_A = \|\varepsilon_0^{CG}\|_A.$$

This together Theorem 9.2.2(a) provides a useful lower bound

(9.2.3) $\quad \|\varepsilon_n^{CG}\|_A^2 = \|\varepsilon_0^{CG}\|_A^2 - \|x_n^{CG} - x_0^{CG}\|_A^2 \geq \|x_m^{CG} - x_0^{CG}\|_A^2 - \|x_n^{CG} - x_0^{CG}\|_A^2,$

which is attained for $m = L$.

Before we finish this part with two illustrating examples, we will discuss an alternative method for approximating the A-norm of the error. It is based on a beautiful connection between the CG method, Gaussian quadrature, and continued fractions and therefore interesting on its own rights.

Recall that the norm in question satisfies as well the following expression (cf. (2.4.9) and Cor. 2.4.6)

(9.2.4) $\qquad \|\varepsilon_n^{CG}\|_A^2 = \|\varepsilon_0^{CG}\|_A^2 - \nu_0 e_1^T J_n^{-1} e_1$

where J_n denotes the Jacobi matrix with respect to $\langle \cdot, \cdot \rangle_{GAL}$ and where

$$\nu_0 = \int_a^d d\sigma(t) = \sum_{j=1}^L \sigma_j^2 = r_0^T r_0.$$

9.2 Approximating the A-Norm of the Error

is the associated zero-order moment, which may be computed in terms of the initial residual r_0. We remark that

$$(9.2.5) \qquad \nu_0 e_1^T J_n^{-1} e_1 = \|x_n^{CG} - x_0^{CG}\|_A^2.$$

Since J_n is available during the CG iteration, the problem of approximating $\|\varepsilon_n^{CG}\|_A$, "reduces" to the approximation of the initial error $\|\varepsilon_0^{CG}\|_A$. Notice, however, that the computation of the $(1,1)$-element of J_n^{-1} involves, at each step, the solution of a linear system

$$J_n y = e_1, \quad \text{with} \quad e_1^T J_n^{-1} e_1 = e_1^T y.$$

A much cheaper way of computing this quantity is based on the connection between orthogonal polynomials, Gaussian quadrature, and continued fractions (cf. Section 2.2) More precisely, in view of Theorem 2.2.6(a), Cor. 2.4.6, and Theorem 2.2.2(b), we have

$$(9.2.6) \qquad \|\varepsilon_0^{CG}\|_A^2 = \langle t^{-1}, 1 \rangle_{GAL} = -\frac{\beta_1^{MO}}{0 - \alpha_1^{MO}} - \frac{\beta_2^{MO}}{0 - \alpha_2^{MO}} - \cdots - \frac{\beta_L^{MO}}{0 - \alpha_L^{MO}},$$

and

$$(9.2.7) \qquad \nu_0 e_1^T J_n^{-1} e_1 = -\frac{\beta_1^{MO}}{0 - \alpha_1^{MO}} - \frac{\beta_2^{MO}}{0 - \alpha_2^{MO}} - \cdots - \frac{\beta_n^{MO}}{0 - \alpha_n^{MO}}.$$

where the α_j^{MO}'s and β_j^{MO}'s are the three-term recurrence coefficients of the monic polynomials

$$(9.2.8) \qquad \begin{aligned} \psi_{-1}^{MO}(t) &:= 0, \quad \psi_0^{MO}(t) = 1, \\ \psi_j^{MO}(t) &= (t - \alpha_j^{MO})\psi_{j-1}^{MO}(t) - \beta_j^{MO}\psi_{j-2}^{MO}(t), \quad j \geq 1, \end{aligned}$$

orthogonal with respect to $\langle \cdot, \cdot \rangle_{GAL}$. Moreover, we introduced the definition

$$(9.2.9) \qquad \beta_1^{MO} := \nu_0.$$

Hence, the approximation of (9.2.4) comes down to the computation of these coefficients and to the evaluation of the continued fractions in (9.2.6) and (9.2.7), respectively.

The latter task may be accomplished by the following algorithm which is due to Gautschi [56, pp. 29]. Before we state this procedure let us reformulate (9.2.7) by means of an equivalence transformation

(9.2.10)
$$C_n := \nu_0 e_1^T J_n^{-1} e_1 = -\frac{\beta_1^{MO}}{0 - \alpha_1^{MO}} - \frac{\beta_2^{MO}}{0 - \alpha_2^{MO}} - \cdots - \frac{\beta_n^{MO}}{0 - \alpha_n^{MO}}$$
$$= -\frac{-\beta_1^{MO}}{\alpha_1^{MO}} + \frac{-\beta_2^{MO}}{\alpha_2^{MO}} + \cdots + \frac{-\beta_n^{MO}}{\alpha_n^{MO}}.$$

Algorithm 9.2.3 *The continued fraction C_n (cf. (9.2.10)) may be computed by the following (forward) recursion*

Set
$$C_0 := 0, \quad C_1 = \frac{\beta_1^{MO}}{\alpha_1^{MO}}, \quad \rho_2 = \left(1 - \frac{\beta_2^{MO}}{\alpha_1^{MO}\alpha_2^{MO}}\right)^{-1}.$$

For $j = 2, 3, \ldots, n$ compute

$$C_j = \rho_j C_{j-1} - (\rho_j - 1)C_{j-2}, \quad \rho_{j+1} = \left(1 - \frac{\beta_{j+1}^{MO}}{\alpha_j^{MO}\alpha_{j+1}^{MO}} \rho_j\right)^{-1}.$$

It remains to devise formulae for the β_j^{MO}'s and α_j^{MO}'s based upon the quantities available during the CG process. Recall, that the residual polynomial for the CG-method looks like

(9.2.11)
$$\psi_{-1}^{OR}(t) := 0, \quad \psi_0^{OR}(t) = 1,$$
$$\gamma_j \psi_j^{OR}(t) = (t - \alpha_j)\psi_{j-1}^{OR}(t) - \beta_j \psi_{j-2}^{OR}(t), \quad j \geq 1.$$

where $\alpha_j = -(\gamma_j + \beta_j)$. In the actual implementation, however, these recurrence coefficients are computed implicitly via (cf. (6.2.6) and (6.2.7))

(9.2.12)
$$\eta_j^{CG} = \frac{1}{\gamma_j} \quad \text{and} \quad \nu_j^{CG} = \frac{\beta_j}{\gamma_{j-1}}.$$

9.2 Approximating the A-Norm of the Error

We proceed as follows. First we derive the monic polynomial ψ_j^{MO} from the residual polynomial ψ_j^{OR} and then we express the obtained coefficients in terms of η_j^{CG} and ν_j^{CG}, respectively. We have with

$$\psi_j^{\text{MO}}(t) = k_j \psi_j^{\text{OR}}(t)$$

the identity

$$\psi_j^{\text{MO}}(t) = \frac{k_j}{\gamma_j}\left((t-\alpha_j)\frac{1}{k_{j-1}}\psi_{j-1}^{\text{MO}}(t) - \frac{\beta_j}{k_{j-2}}\psi_{j-2}^{\text{MO}}(t)\right)$$

$$= (t-\alpha_j)\psi_{j-1}^{\text{MO}}(t) - \beta_j\gamma_{j-1}\psi_{j-2}^{\text{MO}}(t).$$

Combining this with (9.2.11), (9.2.12) and (9.2.9) yields

$$\alpha_j^{\text{MO}} = \alpha_j = -(\gamma_j + \beta_j) = \begin{cases} -\dfrac{1}{\eta_1^{\text{CG}}} & \text{for } j=1 \\[2mm] -\dfrac{1}{\eta_{j-1}^{\text{CG}}\eta_j^{\text{CG}}}(\eta_{j-1}^{\text{CG}} + \eta_j^{\text{CG}}\nu_j^{\text{CG}}) & \text{for } j>1, \end{cases}$$

and

$$\beta_j^{\text{MO}} = \beta_j\gamma_{j-1} = \begin{cases} \nu_1^{\text{CG}} & \text{for } j=1 \\[2mm] \dfrac{\nu_j^{\text{CG}}}{\eta_{j-1}^{\text{CG}}\eta_{j-1}^{\text{CG}}} & \text{for } j>1. \end{cases}$$

Finally, the quantities needed in Algorithm 9.2.3 are easily computed

$$\frac{\beta_1^{\text{MO}}}{\alpha_1^{\text{MO}}} = -\eta_1^{\text{CG}}\nu_1^{\text{CG}}$$

$$\frac{\beta_2^{\text{MO}}}{\alpha_1^{\text{MO}}\alpha_2^{\text{MO}}} = \frac{\eta_2^{\text{CG}}\nu_2^{\text{CG}}}{\eta_1^{\text{CG}} + \eta_2^{\text{CG}}\nu_2^{\text{CG}}}$$

$$\frac{\beta_{j+1}^{\text{MO}}}{\alpha_j^{\text{MO}}\alpha_{j+1}^{\text{MO}}} = \frac{\eta_{j-1}^{\text{CG}}\eta_{j+1}^{\text{CG}}\nu_{j+1}^{\text{CG}}}{(\eta_{j-1}^{\text{CG}} + \eta_j^{\text{CG}}\nu_j^{\text{CG}})(\eta_j^{\text{CG}} + \eta_{j+1}^{\text{CG}}\nu_{j+1}^{\text{CG}})}.$$

This furnishes the procedure cfAerr for approximating the A-norm of the error in terms of continued fractions.

MATLAB Implementation of cfAerr

```
function [C]=cfAerr(nu, eta, tol)
%
% approximates ||x_* - x_0||_A using the
% continued fraction scheme
%
% desired accuracy of the approximation: tol
%

%%initialize
   n=0;
   n_max=length(nu); C=zeros(n_max,1);
   C(1)=0;  C(2)=-eta(1)*nu(1);
   d2=1; rho=1;

%%iterate
   while n < n_max-1 & abs((C(n+2)-C(n+1))/C(n+2)) > tol
      n=n+1;

      d1=d2; d2=eta(n)+eta(n+1)*nu(n+1);
      d=eta(n+1)*nu(n+1)/(d1*d2);
      if n > 1  d=d*eta(n-1);   end;

      rho=1/(1-d*rho);

      C(n+2)=rho*C(n+1)-(rho-1)*C(n);

   end %while

   C=C(1:n+2);

return;
```

Two Examples

The purpose of this subsection is to illustrate that the approximation of the A-norm of the error by either the method outlined in Theorem 9.2.2 or by the continued fraction approach is quite remarkable.

The idea is to substitute the initial error by the best available approximation (cf. (9.2.3))

$$\|\varepsilon_0^{CG}\|_A^2 \approx \|x_m^{CG} - x_0^{CG}\|_A^2,$$

and subsequently to compute

$$\|x_m^{CG} - x_0^{CG}\|_A^2 - \|x_n^{CG} - x_0^{CG}\|_A^2 \approx \|\varepsilon_n^{CG}\|_A^2.$$

9.2 Approximating the A-Norm of the Error

Recall from (9.2.10), (9.2.5), and Theorem 9.2.2(a) that we have various expressions for the approximation

$$C_n = \|x_n^{\text{CG}} - x_0^{\text{CG}}\|_A^2 = \nu_0 e_1^T J_n^{-1} e_1 = \sum_{k=0}^{n-1} |\eta_{k+1}^{\text{CG}}|(r_k^{\text{CG}})^T r_k^{\text{CG}}.$$

So, in theory the scheme based on Theorem 9.2.2 and the scheme based on continued fractions should produce the same results. In fact, in all our experiments we have not noticed a difference between the two approaches.

Here, we will demonstrate the performance of cfAerr, because there exists a roundoff error analysis for this scheme (cf. Golub and Strakoš [70]). We remark that different examples were performed in Fischer and Golub [42].

Example 9.2.4 This is an experiment with a positive definite matrix. Here $A = A(0, 25)$ (cf. (5.1.2)) is the matrix of order $N = 625$, obtained after discretizing Poisson's equation (cf. (5.1.1) with $\tau = 0$) on the unit square. The right hand side was chosen such that $x_* = (1, 1, \ldots, 1)^T$ is the solution and a random vector with components uniformly distributed in $[-1, 1]$ was selected as initial guess x_0. We stopped the iteration after the (relative) norm of the residual was reduced to less than 10^{-14}, which occurred at step m=106. We then took C_{106} as the approximation to the initial error. The graphs of the true error $\log_{10} \|(x_* - x_n^{\text{CG}})\|_A/\|\varepsilon_0\|_2$ and of the approximation $\log_{10} \sqrt{|C_{106} - C_n|}/C_{106}$ are shown in Figure 9.2.5.

Note, that the dashed curve is actually a superposition of the solid curve until about half the machine precision. Let us explain, why suddenly these two curves diverge. A crucial point for the correct error estimation is the convergence of the continued fraction C_n (cf. (9.2.10)) to the initial error $\|\varepsilon_0\|_A$. For this experiment we observed convergence up to machine precision, i.e.,

$$\frac{|C_{n+1} - C_n|}{C_n} < \text{EPS}, \quad \text{for} \quad n \geq n_0 = 78.$$

In fact, we measured

$$\frac{|C_{n+1} - \|\varepsilon_0\|_A^2|}{C_n} < 10^{-14} \quad \text{for} \quad n \geq n_0.$$

In conclusion, $|C_{106} - C_n|$ provides very good approximations to the A-norm of the error, as long as n is not "too close" to the critical parameter n_0, indicated by the dotted curve.

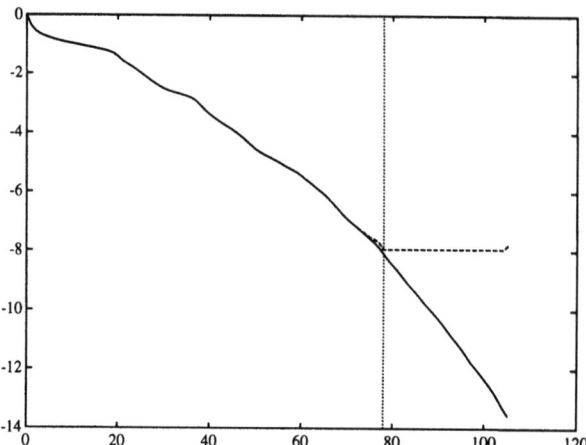

Figure 9.2.5 Approximated and true CG energy norms versus the number of iterations for the matrix $A(0,25)$. Solid curve: $\log_{10} \|(x_* - x_n^{CG})\|_A / \|\varepsilon_0\|_2$; dashed curve: $\log_{10} \sqrt{|C_{106} - C_n|/C_{106}}$; dotted curve: $n \equiv 78$.

This numerical observation was confirmed by roundoff analysis of the underlying (continued fraction) schemes by Golub and Strakoš [70]. Their results are based on work by Greenbaum [73].

Also, we mention that one may as well use one of the approximations C_k, $k \in \{78, 79, \ldots, 105\}$ instead of C_{106}. Any of these substitutions does produce comparably good results.

Example 9.2.6 In this, we attacked an indefinite problem. We applied the CG method and subsequently the continued fraction scheme to the matrix $A = A(350, 25)$ (cf. (5.1.2)), using the same solution and starting guess as in the previous example. This time, CG converged after 245 steps.

Note the divergence of the dashed curve and the solid curve beyond about $n = 175$. In accordance with Example 9.2.4, this is due to the convergence of the continued fractions. This time we observed

$$\frac{|C_{n+1} - C_n|}{C_n} < EPS, \quad \text{for} \quad n \geq n_0 = 181.$$

9.2 Approximating the A-Norm of the Error

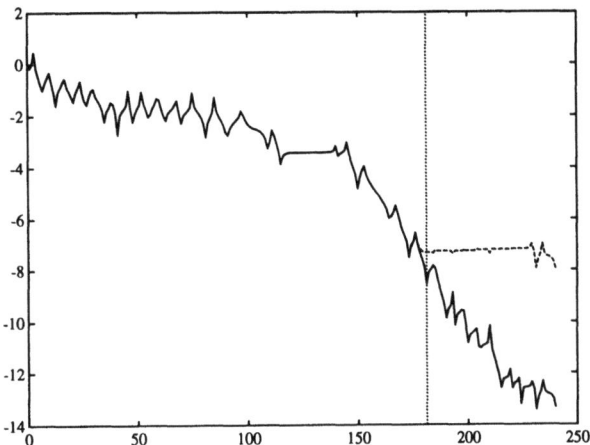

Figure 9.2.7 Approximated and true CG energy "norms" versus the number of iterations for the matrix $A(350, 25)$. Solid curve: $\log_{10} \|(x_* - x_n^{CG})\|_A/\|\varepsilon_0\|_2$; dashed curve: $\log_{10} \sqrt{|C_{245} - C_n|/C_{245}}$; dotted curve: $n \equiv 181$.

General Case

In this section we discuss the approximation of the error for polynomial iteration method defined by

$$\varepsilon_n = p_n(A)\varepsilon_0.$$

The problem of determining the A-norm of the nth error

(9.2.13)
$$\|\varepsilon_n\|_A^2 = r_0^T p_n(A) A^{-1} p_n(A) r_0$$
$$= \int_a^d p_n^2(t) \frac{1}{t} d\sigma(t) =: \int_a^d p_n^2(t) d\tilde{\sigma}(t)$$

is equivalent to the evaluation of an integral with distribution function $\tilde{\sigma}(t)$. Let us assume (for a moment) that we know the orthonormal (†) polynomials $\tilde{\psi}_k$ relative to $\tilde{\sigma}(t)$. If $\tilde{J}_n$ denotes the associated Jacobi matrix, then the integral (9.2.13) may be evaluated by means of Gaussian quadrature (cf. Theorem 2.2.1)

$$\|\varepsilon_n\|_A^2 = \tilde{\nu}_0 e_1^T p_n(\tilde{J}_n) p_n(\tilde{J}_n) e_1,$$

(†) the assumption of orthonormality is more convenient but not necessary.

where

$$\tilde{\nu}_0 = \int_a^d d\tilde{\sigma}(t) = \|\varepsilon_0\|_A^2$$

denotes the zero-order moment, which turns out to be the square of the norm of the initial error. Hence, the approximation of $\|\varepsilon_n\|_A$ "reduces" to the computation of $\tilde{J}_n$ and $\tilde{\nu}_0$, respectively.

There are essentially two ways of deriving $\tilde{J}_n$. Both approaches assume that the (orthonormal) polynomials ψ_k with respect to $\sigma(t)$ are given. So, let us assume that the ψ_k's are available (how to actually compute these polynomials will be described at the end of this section).

The first approach is based upon modified moments connected to $\tilde{\sigma}(t)$. It is well-known (compare Section 2.2) that the first $2n+1$ modified moments

$$(9.2.14) \qquad \tilde{\nu}_j = \tilde{\nu}(\psi_j^{\text{MO}}; \tilde{\sigma}) = \int_a^d \frac{\psi_j^{\text{MO}}(t)}{t} d\sigma(t), \quad j = 0, 1, \ldots, 2n,$$

determine the desired Jacobi matrix $\tilde{J}_n$. Here, $\{\psi_j^{\text{MO}}\}_{j=0}^{2n}$ is the set of monic orthogonal polynomials with respect to $\sigma(t)$. In other words, these moments are the minimal solution (with parameter $\xi = 0$), as analyzed in Theorem 2.2.7. Hence, the direct computation of (9.2.14) is not an easy problem. Fortunately, to accomplish this task, Gautschi [56, pp. 37] devised a stable (backward recurrence) algorithm based on continued fractions.

Let $M > 2n$ be given and let α_j^{MO} and β_j^{MO} denote the three-term recurrence coefficients of ψ_j^{MO} (cf. (9.2.8)). Consider the following backward recurrence

$$h_M^{(M)} = 0, \quad h_{m-1}^{(M)} = \frac{\beta_{m+1}^{\text{MO}}}{h_m^{(M)} - \alpha_{m+1}^{\text{MO}}} \quad m = M, M-1, \ldots, 0,$$

$$\hat{\nu}_{-1}^{(M)} = -1, \quad \hat{\nu}_j^{(M)} = h_{j-1}^{(M)} \hat{\nu}_{j-1}^{(M)} \quad j = 0, 1, \ldots, 2n.$$

Then, as was proven by Gautschi [56, p. 39],

$$\lim_{M \to \infty} \hat{\nu}_j^{(M)} = \hat{\nu}_j, \quad j = 0, 1, \ldots, 2n.$$

Our experience (cf. Fischer and Golub [41], [42]) with this scheme is quite satisfactory. However, a remark is appropriate here. For this algorithm to be effective, one needs "good" estimates for the startindex M. On the other hand, the scheme is cheap in terms of operation count. So, a restart with an increased index M is affordable.

9.2 Approximating the A-Norm of the Error

Next, we will briefly explain the second approach for generating $\hat{J}_n$ (for details we refer to Fischer and Golub [41] and to Gautschi [59]). The idea is to view

$$\hat{\sigma}(t) = \frac{1}{t}\sigma(t)$$

as a *modification* of $\sigma(t)$. In this light, the orthogonal polynomials with respect to $\sigma(t) = t\hat{\sigma}(t)$ may be viewed as kernel polynomials (cf. Theorem 2.5.1) with respect to $t\hat{\sigma}(t)$. To devise a scheme for the computation of $\hat{J}_n$, the trick is to assume that we already know the desired matrix. Then, based on this information, to "artificially generate" the kernel polynomials and finally to "invert" the latter process.

The resulting scheme often works although there are instances where it becomes unstable. The stability properties of the process need further investigation. On the other hand, there is no startindex needed for this approach, as for the previous scheme.

It remains to show how to compute the orthonormal polynomials ψ_j with respect to $\sigma(t)$.

Recall, that in the CG case, these polynomials are essentially the residual polynomials $\psi_j(t) = \psi_j(0)\psi_j^{\text{OR}}(t)$. For the SYMMLQ scheme and the MINRES scheme, respectively, the situation is just as convenient. Here, in both implementations, the "Lanczos part" directly provides the three-term recurrence coefficients of the ψ_j's (compare Section 4.1).

Other polynomial iteration methods need some further investigation. Let us discuss a general technique which is applicable to any of these methods. Again, we "parameterize" the desired polynomials in terms of modified moments. To this end consider the polynomial iteration scheme defined by

$$r_n = p_n(A)r_0,$$

and observe (compare Dahlquist, Eisenstat and Golub [27])

$$r_j^T r_0 = r_0^T p_j(A) r_0 = \sum_{k=1}^{L} \sigma_k^2 p_j(\lambda_k) = \int_a^d p_j(t)d\sigma(t) = \nu(p_j;\sigma) =: \nu_j,$$

where we have used the representation (1.1.6) of the start residual r_0. Hence, after $2n$ steps of the polynomial iteration method we have accumulated enough information to compute the first n orthogonal polynomials with respect to $\sigma(t)$. Obviously, it would be advantageous to have these polynomials after only n iteration steps. In fact, due to the special properties of Chebyshev polynomials, this can be done for the Chebyshev iteration (compare Golub and Kent [69]). However, it is worth noticing, that the price paid for the computation of one modified moment ν_j is one inner product $r_j^T r_0$.

Lower and Upper Bounds

It is also possible to derive lower and upper bounds for $\|\varepsilon_n\|_A$. This technique has been described in papers by Dahlquist,Eisenstat and Golub [27], Golub [67], and Dahlquist, Golub and Nash [26]

Their scheme exploits the connection between modified moments and *Gauß-Radau* quadrature rules. To be effective, however, the method requires the knowledge of upper and lower bounds for the spectrum of the system matrix A.

10 Bibliography

[1] N. I. ACHIESER, *Über einige Funktionen, welche in zwei gegebenen Intervallen am wenigsten von Null abweichen I. Teil*, Bull. Acad. Sci. URSS S. VII, 9 (1932), pp. 1163–1202.

[2] ———, *Über eine Eigenschaft der "elliptischen" Polynome*, Comm. Kharkov Math. Soc., 9 (1934), pp. 3–8.

[3] ———, *Elements of the Theory of Elliptic Functions*, Amer. Math. Soc., Providence, R. I., 1990.

[4] L. AHLFORS, *Complex analysis*, McGraw - Hill, New York, second ed., 1966.

[5] R. S. ANDERSSEN AND G. H. GOLUB, *Richardson's non-stationary matrix iterative procedure*, rep. stan-cs-72-304, Computer Science Dept., Stanford University, 1972.

[6] S. F. ASHBY, T. A. MANTEUFFEL, AND P. E. SAYLOR, *A taxonomy for conjugate gradient methods*, SIAM J. Numer. Anal., 27 (1990), pp. 1542–1568.

[7] B. ATLESTAM, *Tschebycheff-polynomials for sets consisting of two disjoint intervals with application to convergence estimates for the conjugate gradient method*, Res. Rep. 77.06, Dept. Computer Science, Chalmers University of Technology and the University of Göteborg, 1977.

[8] O. AXELSSON AND V. BARKER, *Finite element solution of boundary value problems: theory and computation*, Academic, New York, 1984.

[9] M. BERCOVIER AND O. PIRONNEAU, *Error estimates for finite element solution of the Stokes problem in primitive variables*, Numer. Math., 33 (1979), pp. 211–224.

[10] D. BRAESS, *Finite Elemente*, Springer, Berlin, Heidelberg, New York, 1992.

[11] F. BREZZI, *On the existence, uniqueness and approximation of saddle-point problems arising from Lagrangian multipliers*, RAIRO Anal. Numer., 8 (1974), pp. 129–151.

[12] F. BREZZI AND M. FORTIN, *Mixed and hybrid finite element methods*, Springer, New York, 1991.

[13] K. W. BRODLIE, *A review of methods for curve and function drawing*, in Mathematical Methods in Computer Graphics and Design, K. W. Brodlie, ed., London, 1980, Academic, pp. 1–37.

[14] P. F. BYRD AND M. D. FRIEDMAN, *Handbook of Elliptic Integrals for Engineers and Scientists*, Springer, Berlin, Göttingen, Heidelberg, 1954.

[15] D. CALVETTI, L. REICHEL, AND Q. ZHANG, *Conjugate gradient algorithms for symmetric inconsistent linear systems*, in Proceedings of the Cornelius Lanczos International Centenary Conference, J. Brown, M. Chu, D. Ellison, and R. Plemmons, eds., Philadelphia, 1994, SIAM, pp. 267–272.

[16] B. CARLSON, *Hidden symmetries of special functions*, SIAM Rev., 12 (1970), pp. 332–345.

[17] ——, *Special functions of applied mathematics*, Academic, New York, 1977.

[18] R. CHANDRA, *Conjugate gradient methods for partial differential equations*, PhD thesis, Yale university, 1978. Res. Rep. 129.

[19] R. CHANDRA, S. C. EISENSTAT, AND M. H. SCHULTZ, *The modified conjugate residual method for partial differential equations*, in Advances in computer methods for partial differential equations II, R. Vichnevetsky, ed., AICA, 1977, IMACS, pp. 13–19.

[20] P. L. CHEBYSHEV, *Oeuvre I*, L'Academie Imperiale de Science, St. Petersbourg, 1899. A. Markoff and N. Sonin, eds.

[21] E. CHENEY, *Introduction to approximation theory*, McGraw-Hill, New York, 1966.

[22] T. S. CHIHARA, *An introduction to orthogonal polynomials*, Gordon and Breach, New York, London, Paris, 1978.

[23] R. COURANT AND D. HILBERT, *Methoden der Mathematischen Physik I*, Springer, Berlin, 1924.

[24] E. J. CRAIG, *The N-step iteration procedures*, J. Math. Phys., 34 (1955), pp. 64–73.

[25] J. CULLUM AND A. GREENBAUM, *Residual relationships with three pairs of iterative algorithms for solving Ax=b*, Research Report RC 18672 (81716) 1/27/93, IBM Research Divion, T.J. Watson Research Center, 1993.

[26] G. DAHLQUIST, G. H. GOLUB, AND S. NASH, *Bounds for the error in linear systems*, in Proceedings of the Workshop on Semi-Infinite Programming, R. Hettich, ed., Springer, 1978, pp. 154–172.

[27] G. G. DAHLQUIST, S. C. EISENSTAT, AND G. H. GOLUB, *Bounds for the error of linear systems of equations using the theory of modified moments*, J. Math. Anal. Appl., 37 (1972), pp. 151–166.

[28] P. J. DAVIS, *Interpolation & Approximation*, Blaisdell, Waltham, Massachusetts, 1963.

[29] C. DE BOOR, *A Practical Guide to Splines*, Springer, New York, 1978.

[30] C. DE BOOR AND J. R. RICE, *Extremal polynomials with application to Richardsons iteration for indefinite linear systems*, SIAM J. Sci. Stat. Comput., 3 (1982), pp. 47–57.

[31] P. DEUFLHARD, *A study of Lanczos-type iterations for symmentric indefinite linear systems*, Preprint SC 93-6, Konrad Zuse Zentrum, Berlin, 1993.

[32] ———, *Cascadic conjugate gradient methods for elliptic partial differential equations. algorithm and numerical results*, in Proceedings of the seventh international conference on domain decomposition, D. Keyes and J. Xu, eds., Providence, 1994, American Mathematical Society, pp. 29–42.

[33] M. EIERMANN, X. LI, AND R. S. VARGA, *On hybrid semi-iterative methods*, SIAM J. Numer. Anal., 26 (1989), pp. 152–168.

[34] M. EIERMANN AND W. NIETHAMMER, *On the construction of semi-iterative methods*, SIAM J. Numer. Anal., 20 (1983), pp. 1153–1160.

[35] M. EIERMANN, W. NIETHAMMER, AND R. S. VARGA, *A study of semiiterative methods for nonsymmetric systems of linear equation*, Numer. Math., 47 (1985), pp. 505–533.

[36] H. ELMAN, *Multigrid and Krylov subspace methods for the discrete Stokes equations*, Report UMIACS-TR-94-76, University of Maryland, 1994. To appear in Int. J. Numer. Meth. Fluids, 1995.

[37] B. FISCHER, *Chebyshev polynomials for disjoint compact sets*, Constr. Approx., 8 (1992), pp. 309–329.

[38] B. FISCHER AND R. W. FREUND, *An inner product-free conjugate gradient-like algorithm for Hermitian positive definite systems*, in Proceedings of the Cornelius, Lanczos International Centenary Conference, J. Brown, M. Chu, D. Ellison, and R. Plemmons, eds., SIAM, 1994, pp. 288–290.

[39] ———, *On adaptive weighted polynomial preconditioning for Hermitian positive definite matrices*, SIAM J. Sci. Comput., 15 (1994), pp. 408–426.

[40] B. FISCHER AND G. H. GOLUB, *On generating polynomials which are orthogonal over several intervals*, Math. Comp., 56 (1991), pp. 711–730.

[41] ———, *How to generate unknown orthogonal polynomials out of known orthogonal polynomials*, J. Comp. Appl. Math., 43 (1992), pp. 99–115.

[42] ———, *On the error computation for polynomial based iteration methods*, in Recent Advances in Iterative Methods, A. G. G. Golub and M. Luskin, eds., Springer, 1993, pp. 59–69.

[43] B. FISCHER, M. HANKE, AND M. HOCHBRUCK, *A note on conjugate - gradient type methods for indefinite and/or inconsistent linear systems*, technical report, Medical University of Lübeck, Institute of Mathematics, 1994. To appear in Numer. Alg.

[44] B. FISCHER AND J. MODERSITZKI, *An algorithm for complex linear approximation based on semi-infinite programming*, Numer. Algor., 5 (1993), pp. 287–299.

[45] B. FISCHER, A. RAMAGE, D. SILVESTER, AND A. WATHEN, *Minimum residual methods for augmented systems*, Technical Report A-95-19, Medical University of Lübeck, Institute of Mathematics, 1995.

[46] R. FLETCHER, *Conjugate gradient methods for indefinite systems*, in Numerical Analysis–Dundee 1975, G. Watson, ed., Heidelberg, 1976, Springer, pp. 73–89. Lecture Notes in Mathematics, volume 506.

[47] R. W. FREUND, *Über einige CG-ähnliche Verfahren zur Lösung linearer Gleichungssysteme*, PhD thesis, Universität Würzburg, FRG, 1983. In German.

[48] ———, *Pseudo-Ritz values for indefinite Hermitian matrices*, Technical Report TR 89.33, RIACS, 1989.

[49] ———, *On conjugate gradient type methods and polynomial preconditioners for a class of complex non-Hermitian matrices*, Numer. Math., 57 (1990), pp. 285–312.

[50] ———, *On polynomial preconditioning and asymptotic convergence factors for indefinite Hermitian matrices*, Linear Alg. Appl., 154–156 (1991), pp. 259–288.

[51] ———, *Quasi-kernel polynomials and their use in non-Hermitian matrix iterations*, J. Comp. Appl. Math., 43 (1992), pp. 135–158.

[52] R. W. FREUND, G. H. GOLUB, AND N. M. NACHTIGAL, *Iterative solutions of linear systems*, Acta Numerica, (1992), pp. 1–44.

[53] V. M. FRIDMAN, *The method of minimum iterations with minimium errors for a system of linear algebraic equations with symmetrical matrix*, USSR Comput. Math. and Math. Phys., 2 (1963), pp. 362–363.

[54] F. N. FRITSCH AND J. BUTLAND, *A method for constructing local monotone piecewise cubic interpolation*, SIAM J. Sci. Stat. Comput., 5 (1984), pp. 300–304.

[55] F. N. FRITSCH AND R. E. CARLSON, *Monotone piecewise cubic interpolation*, SIAM J. Numer. Anal., 17 (1980), pp. 238–246.

[56] W. GAUTSCHI, *Computational aspects of three-term recurrence relations*, SIAM Rev., 9 (1967), pp. 24–82.

[57] ———, *Zur Numerik rekurrenter Relationen*, Computing, 9 (1972), pp. 107–126.

[58] ———, *A survey of Gauss-Christoffel quadrature formulae*, in E. B. Christoffel: The influence of his work in mathematics and the physical sciences, P. Butzer and F. Fehér, eds., Basel, 1981, Birkhäuser, pp. 72–147.

[59] ———, *An algorithmic implementation of the generalized Christoffel theorem*, in Numerical Integration, G. Hämmerlin, ed., Basel, 1982, Birkhäuser, pp. 89–106. Internat. Ser. Numer. Math., v. 57.

[60] ——, *On generating orthogonal polynomials*, SIAM J. Sci. Stat. Comput., 3 (1982), pp. 289–317.

[61] ——, *On the sensitivity of orthogonal polynomials to perturbations in the moments*, Numer. Math., 48 (1986), pp. 369–382.

[62] ——, *Orthogonality – conventional and unconventional – in numerical analysis*, in Computation and Control, K. Bowers and J. Lund, eds., Boston, Basel, Berlin, 1989, Birkhäuser, pp. 63–95. Proceedings of the Bozeman Conference.

[63] J. GERONIMO AND W. VAN ASSCHE, *Orthogonal polynomials on several intervals via a polynomial mapping*, Trans. Amer. Math. Soc., 308 (1988), pp. 559–581.

[64] V. GIRAULT AND P.-A. RAVIART, *Finite element methods for Navier-Stokes equations, theory and algorithms*, Springer, Berlin, Heidelberg, 1986.

[65] G. GOLUB AND R. VARGA, *Chebychev semi-iterative methods, successive overrelaxation iterative methods, and second order Richardson iterative methods*, Numer. Math., 3 (1961), pp. 147–168. Part I and Part II.

[66] G. GOLUB AND J. WELSCH, *Calculation of Gauss quadrature rules*, Math. Comp., 23 (1969), pp. 221–230.

[67] G. H. GOLUB, *Some modified matrix eigenvalue problems*, SIAM Rev., 15 (1973), pp. 318–334.

[68] G. H. GOLUB AND W. KAHAN, *Calculating the singular values and pseudoinverse of a matrix*, SIAM J. Numer. Anal., 2 (1965), pp. 205–224.

[69] G. H. GOLUB AND M. D. KENT, *Estimates of eigenvalues for iterative methods*, Math. Comp., 53 (89), pp. 619–626.

[70] G. H. GOLUB AND Z. STRAKOŠ, *Estimates in quadratic formulas*, Numer. Algor., 8 (1994), pp. 241–268.

[71] G. H. GOLUB AND C. F. VAN LOAN, *Matrix Computations*, The Johns Hopkins University Press, Baltimore, second ed., 1989.

[72] J. F. GRCAR, *Analyses of the Lanczos algorithm and of the approximation problem in Richardson's method*, PhD thesis, Dept. Computer

Science, University of Illinois at Urbana-Champaign, 1981. Rept. No. UIUCDCS-R-81-1074.

[73] A. GREENBAUM, *Behavior of slightly perturbed Lanczos and conjugate-gradient recurrences*, Linear Alg. Appl., 113 (1989), pp. 7–63.

[74] M. D. GUNZBURGER, *Finite element methods for viscous incompressible flows: a guide to theory, practice, and algorithms*, Academic, London, 1989.

[75] M. GUTKNECHT, *Changing the norm in conjugate gradient type algorithms*, SIAM J. Numer. Anal., 30 (1993), pp. 40–56.

[76] W. HACKBUSCH, *Theorie und Numerik elliptischer Differentialgleichungen*, Teubner, Stuttgart, 1986.

[77] ——, *Elliptic Differential Equations*, Springer, 1992.

[78] L. HELMS, *Introduction to potential theory*, Wiley, New York, 1969. Pure and Applied Mathematics, vol. XXII.

[79] P. HENRICI, *Applied and computational complex analysis, Vol. 1*, Wiley, New York, 1974.

[80] ——, *Applied and computational complex analysis, Vol. 3*, Wiley, New York, 1986.

[81] M. R. HESTENES AND E. STIEFEL, *Methods of conjugate gradients for solving linear systems*, J. Res. Nat. Bur. Stand., 49 (1952), pp. 409–436.

[82] E. HILLE, *Analytic Function Theory*, Ginn and Co., Boston, 1962. Introduction to Higher Mathematics, vol. II.

[83] A. S. HOUSEHOLDER, *The theory of matrices in numerical analysis*, Blaisdell, New York, 1964.

[84] C. JOHNSON, *Numerical Solution of partial differential equations by the finite element methd*, Cambridge University Press, Cambridge, 5. printing ed., 1994.

[85] S. KARLIN AND L. S. SHAPLEY, *Geometry of Moment Spaces*, Amer. Math. Soc., Providence, 1953. Memoirs of the Amer. Math. Soc., 12.

[86] H. KOBER, *Dictionary of conformal representations*, Dover, New York, 1957.

[87] C. LANCZOS, *An iteration method for the solution of the eigenvalue problem of linear differential and integral operators*, J. Res. Natl. Bur. Stand., 45 (1950), pp. 255–282.

[88] V. I. LEBEDEV, *An iteration method for the solution of operator equations with their spectrum lying on several intervals*, Zh. Vych. Mat. i Fiz., 9 (1969), pp. 1247–1252. in Russian; an English translation has appeared in [5].

[89] T. MANTEUFFEL AND J. OTTO, *On the roots of the orthogonal polynomials and residual polynomials associated with a conjugate gradient method*, J. Num. Lin. Alg. Appl., 1 (1994), pp. 449–477.

[90] W. MARKOFF, *Über Polynome, die in einem gegebenen Intervalle möglichst wenig von Null abweichen*, Math. Ann., 77 (1916), pp. 213–258.

[91] G. MEINARDUS, *Approximation of Functions: Theory and Numerical Methods*, Springer, Berlin, Heidelberg, New York, 1967.

[92] J. MODERSITZKI, *Polynomiale Iterationsverfahren zur Lösung indefiniter linearer Gleichungssysteme mit optimalen Iterierten und kurzer Rekursion*, phd thesis, Institute of Applied Mathematics, University of Hamburg, 1995. In german.

[93] C. PAIGE, B. PARLETT, AND H. VAN DER VORST, *Approximate solutions and eigenvalue bounds from Krylov subspaces*, Numerical Linear Algebra with Applications, 2 (1995), pp. 115–135.

[94] C. C. PAIGE, *Bidiagonalization of matrices and solution of linear systems*, SIAM J. Numer. Anal., 11 (1974), pp. 197–209.

[95] C. C. PAIGE AND M. A. SAUNDERS, *Solution of sparse indefinite systems of linear equations*, SIAM J. Numer. Anal., 12 (1975), pp. 617–629.

[96] ———, *LSQR: an algorithm for sparse linear equations and sparse least squares*, ACM Trans. Math. Software, 8 (1982), pp. 43–71.

[97] B. N. PARLETT, *The Symmetric Eigenvalue Problem*, Prentice-Hall, Englewood Cliffs, 1980.

[98] F. PEHERSTORFER, *Extremalpolynome in der L^1- und L^2-Norm auf Zwei disjunkten Intervallen*, in Approximation Theory and Functional Analysis, P. L. Butzer, R. L. Stens, and B. Sz.-Nagy, eds., Basel, 1984, Birkhäuser, pp. 269–280. ISNM. 65.

[99] ——, *Gauss-Tchebycheff quadrature formulas*, Numer. Math., 58 (1990), pp. 273–286.

[100] ——, *Orthogonal- and Chebyshev polynomials on two intervals*, Acta Math. Hung., 55 (1990), pp. 245–278.

[101] O. PERRON, *Die Lehre von den Kettenbrüchen, Vol. II*, Teubner, Stuttgart, third ed., 1957.

[102] R. PEYRET AND T. TAYLOR, *Computational methods for fluid flow*, Springer Series in Comp. Physics, 1983.

[103] C. POSSE, *Sur les quadratures*, Nouv. Ann. Math., 14 (1875), pp. 49–62.

[104] J. K. REID, *On the method of conjugate gradients for the solution of large sparse systems of equations*, in Large sparse sets of linear equations, J. K. Reid, ed., London, 1971, Academic, pp. 231–254.

[105] W. P. REINHARDT, L^2 *discretization of atomic and molecular electronic continua: moment, quadrature and J-matrix techniques*, Comput. Phys. Comm., 17 (1979), pp. 1–21.

[106] H. RUTISHAUSER, *Theory of gradient methods*, in Refined iterative methods for computation of the solution and the eigenvalues of self-adjoint boundary value problems, M. Engeli, T. Ginsburg, H. Rutishauser, and E. Stiefel, eds., Basel, Stuttgart, 1959, Birkhäuser. Mitteilungen aus dem Institut für angewandte Mathematik an der ETH Zürich 8.

[107] Y. SAAD, *Iterative solution of indefinite symmetric linear systems by methods using orthogonal polynomials over two disjoint intervals*, SIAM J. Numer. Anal., 20 (1983), pp. 784–810.

[108] Y. SAAD AND M. H. SCHULTZ, *GMRES: A generalized minimal residual algorithm for solving linear systems*, SIAM J. Sci. Stat. Comput., 7 (1986), pp. 856–869.

[109] R. A. SACK AND A. F. DONOVAN, *An algorithm for Gaussian quadrature given modified moments*, Numer. Math., 18 (1971/72), pp. 465–478.

[110] F. W. SAUER, *A FORTRAN program for the calculation of an extremal polynomial*, ACM Trans. Math. Software, 9 (1983), pp. 381–383. Algorithm 604.

[111] W. SCHÖNAUER, *Scientific computing on vector computers*, North Holland, Amsterdam, New York, Oxford, Tokyo, 1987.

[112] I. B. SKA, *The finite element method with Lagrangian multipliers*, Numer. Math., 20 (1973), pp. 179–192.

[113] G. SLEIJPEN, H. VAN DER VORST, AND D. FOKKEMA, *BiCGstab(ℓ) and other hybrid Bi-CG methods*, Numer. Algor., 7 (1994), pp. 75–109.

[114] V. I. SMIRNOV, *A course of higher mathematics, Vol. 5*, Addison-Wesley, Reading, Massachusetts, 1964.

[115] E. L. STIEFEL, *Relaxiationsmethoden bester Strategie zur Lösung linearer Gleichungssysteme*, Comm. Math. Helv., 29 (1955), pp. 157–179.

[116] ———, *Kernel polynomials in linear algebra and their numerical applications*, U.S. Nat. Bur. Standards, Appl. Math. Ser., 49 (1958), pp. 1–22.

[117] T. J. STIELTJES, *Sur certaines inégalités dues à M. P. Tchebychef*, in Oeuvres Complètes, Vol. 2, pp. 586–593. Article rédigé d'après un manuscript inédit.

[118] J. STOER, *Solution of large linear systems of equations by conjugate gradient type methods*, in Mathematical programming – the state of the art, A. Bachem, M. Grötschel, and B. Korte, eds., Berlin, Heidelberg, New York, Tokyo, 1983, Springer, pp. 540–565.

[119] J. STOER AND R. W. FREUND, *On the solution of large linear systems of equations by conjugate gradient algorithms*, in Computer Methods in Applied Science and Engineering – V, R. Glowinski and J. Lions, eds., Amsterdam, 1982, North Holland, pp. 35–53.

[120] G. W. STRUBLE, *Orthogonal polynomials: variable-signed weight functions*, Numer. Math., 5 (1963), pp. 88–94.

[121] G. SZEGÖ, *Über orthogonale Polynome, die zu einer gegebenen Kurve der komplexen Ebene gehören*, Math. Zeitschr., 9 (1921), pp. 218–270.

[122] ———, *Orthogonal polynomials*, AMS Colloquium Publications XXIII, American Mathematical Society, New York, revised ed., 1959.

[123] D. B. SZYLD AND O. B. WIDLUND, *Variational analysis of some conjugate gradient methods*, East–West J. Numer. Math, 1 (1992), pp. 1–25.

Bibliography 273

[124] A. VAN DER SLUIS AND H. A. VAN DER VORST, *The rate of convergence of conjugate gradients*, Numer. Math., 48 (1986), pp. 543–560.

[125] ——, *The convergence behavior of Ritz values in the presence of close eigenvalues*, Linear Alg. Appl., 88/89 (1987), pp. 651–694.

[126] R. S. VARGA, *Matrix iterative analysis*, Englewood Cliffs, Prentice Hall, 1962.

[127] J. L. WALSH, *Interpolation and approximation by rational functions in the complex domain*, Amer. Math. Soc., Providence, R. I., third ed., 1960.

[128] A. WATHEN, *Realistic eigenvalue bounds for the Galerkin mass matrix*, IMA J. Numer. Anal., 7 (1987), pp. 449–457.

[129] A. WATHEN, B. FISCHER, AND D. SILVESTER, *The convergence rate of the minimal residual method for the Stokes problem*, Numer. Math., 71 (1995), pp. 121–134.

[130] A. WATHEN AND D. SILVESTER, *Fast iterative solution of stabilised Stokes systems part II: using simple diagonal preconditioners*, SIAM J. Numer. Anal., 30 (1993), pp. 630–649.

[131] R. WEISS, *Convergence behavior of generalized conjugate gradient methods*, PhD thesis, University of Karlsruhe, 1990.

[132] J. C. WHEELER, *Modified moments and Gaussian quadrature*, Rocky Mt. J. Math., 4 (1974), pp. 287–296.

[133] E. T. WHITTAKER AND G. N. WATSON, *A course of modern analysis*, Cambridge University Press, London, fourth ed., 1965. Reprinted.

[134] H. WIDOM, *Extremal polynomials associated with a system of curves in the complex plane*, Adv. in Math, 3 (1969), pp. 127–232.

[135] H. WILF, *Mathematics for the Physical Sciences*, Wiley, New York, 1962.

[136] J. H. WILKINSON, *The Algebraic Eigenvalue Problem*, Clarendon Press, Oxford, 1965.

[137] J. WIMP, *Computation with recurrence relations*, Pitman, Boston, London, Melbourne, 1984.

[138] L. ZHOU AND H. F. WALKER, *Residual smoothing techniques for iterative methods*, SIAM J. Sci. Comput., 15 (1994), pp. 297–312.

11 Notation

Matrices and Vectors

A	$N \times N$ symmetric matrix
λ_j	eigenvalue of A, $\lambda_1 < \lambda_2 < \cdots < \lambda_L$
z_j	eigenvector of A, $Az_j = \lambda_j z_j$
f	right hand side vector
x_*	solution of $Ax = f$
x_n	nth iterate
r_n	residual at step n, $r_n = f - Ax_n$
ε_n	error at step n, $e_n = x_* - x_n$
L	grade of r_0 w.r.t. A, cf. (1.1.5)
$\mathcal{K}_n(A; r_0)$	Krylov subspace, cf. (1.1.4)
$A(\tau, m)$	matrix generated by the model problem, cf. (5.1.2)
e_n	nth unit vector
I_N	$N \times N$ identity matrix
J_n	Jacobi matrix, cf. (2.1.24)
$\hat{J}_n$	"extended" Jacobi matrix, cf. (2.1.25)
Q_n, R_n	QR factors of $\hat{J}_n$, cf. Section 2.3
G_n	Givens rotation defining Q_n, cf. Section 2.3
J_n^{NE}	bidiagonal matrix, cf. (6.8.5)
$\hat{J}_n^{\text{NE}}$	"extended" bidiagonal matrix, cf. (6.8.6)
$Q_n^{\text{NE}}, R_n^{\text{NE}}$	QR factors of $\hat{J}_n^{\text{NE}}$, cf. (6.8.10), (6.8.11)
G_n^{NE}	Givens rotation defining Q_n^{NE}, cf. Lemma 6.8.1

Notation

Orthogonal Polynomials

Π_{n-1}	space of polynomials of degree $\leq n-1$
$\langle \cdot, \cdot \rangle, \|\cdot\|$	inner product on Π_{n-1} or $\mathcal{K}_n$
$\langle \cdot, \cdot \rangle_{\text{GAL}}$	inner product induced by r_0 and A, $\langle p, q \rangle_{\text{GAL}} = r_0^T p(A) q(A) r_0$
$\langle \cdot, \cdot \rangle_n$	inner product induced by the Gaussian quadrature rule, cf. (2.2.4)
$\langle \cdot, \cdot \rangle_S$	inner product induced by a spline, cf. (5.3.5)
ψ_n	orthonormal polynomial
α_j, β_j	three-term recurrence coefficients, cf. (2.1.17)
$\tilde{\psi}_n$	polynomial of the second kind, cf. (2.2.15)
ψ_n^{MO}	monic orthogonal polynomial
ψ_n^{OR}	scaled orthogonal polynomial, $\psi_n^{\text{OR}}(0) = 1$, cf. Section 2.4
ψ_n^S	orthonormal polynomial w.r.t $\langle \cdot, \cdot \rangle_S$, cf. (5.3.6)
$K_n(t; \xi)$	kernel polynomial, cf. Section 2.5
ψ_n^{MR}	scaled kernel polynomial, $\psi_n^{\text{MR}}(0) = 1$, cf. Theorem 2.5.1
$K_n^{\text{H}}(t; \xi)$	Hermite kernel polynomial, cf. Section 2.6
ψ_n^{ME}	scaled Hermite kernel polynomial, $\psi_n^{\text{ME}}(0) = 1$, cf. Theorem 2.6.1
$\sigma(t)$	distribution function, general
$\tau(t)$	distribution function, Gaussian quadrature, cf. (2.2.5)
$\tau_j^{(n)}$	weight of the Gaussian quadrature, cf. (2.2.1)
$\theta_j^{(n)}$	zero of ψ_n, cf. Theorem 2.1.5
ν_0	zero-order moment, cf. (2.1.21)

Polynomial Iteration Methods

$x_n^{\text{CG}}, r_n^{\text{CG}}$	CG implementation, $r_n^{\text{CG}} = \psi_n^{\text{OR}}(A)r_0 = f - Ax_n^{\text{CG}}$, cf. Theorem 6.2.2
$x_n^{\text{CR}}, r_n^{\text{CR}}$	CR implementation, $r_n^{\text{CR}} = \psi_n^{\text{MR}}(A)r_0 = f - Ax_n^{\text{CR}}$, cf. Theorem 6.2.3
$x_n^{\text{STOD}}, r_n^{\text{STOD}}$	STOD implementation, $r_n^{\text{STOD}} = \psi_n^{\text{ME}}(A)r_0 = f - Ax_n^{\text{STOD}}$, cf. Theorem 6.4.1
$x_n^{\text{MCR}}, r_n^{\text{MCR}}$	MCR implementation, $r_n^{\text{MCR}} = \psi_n^{\text{MR}}(A)r_0 = f - Ax_n^{\text{MCR}}$, cf. Theorem 6.4.2
$x_n^{\text{OR}}, r_n^{\text{OR}}$	OR approach, $r_n^{\text{OR}} = \psi_n^{\text{OR}}(A)r_0 = f - Ax_n^{\text{OR}}$, cf. (6.5.3)
$x_n^{\text{ME}}, r_n^{\text{ME}}$	SYMMLQ implementation, $r_n^{\text{ME}} = \psi_n^{\text{ME}}(A)r_0 = f - Ax_n^{\text{ME}}$, cf. Theorem 6.5.1
$x_n^{\text{MR}}, r_n^{\text{MR}}$	MINRES implementation, $r_n^{\text{MR}} = \psi_n^{\text{MR}}(A)r_0 = f - Ax_n^{\text{MR}}$, cf. Theorem 6.5.2
$x_n^{\text{MtE}}, \varepsilon_n^{\text{MtE}}$	Mte implementation, $\varepsilon_n^{\text{MtE}} = x_* - x_n^{\text{MtE}}$, cf. Section 6.7
$x_n^{\text{LSQR}}, r_n^{\text{LSQR}}$	LSQR implementation, $r_n^{\text{LSQR}} = f - Ax_n^{\text{LSQR}}$, cf. Theorem 6.8.2
$x_n^{\text{CRAIG}}, r_n^{\text{CRAIG}}$	CRAIG implementation, $r_n^{\text{CRAIG}} = f - Ax_n^{\text{CRAIG}}$, cf. Theorem 6.8.3
$x_n^{\text{CI}}, r_n^{\text{CI}}$	Chebyshev iteration, $r_n^{\text{CI}} = \mathcal{P}_n(A; \Omega, 0)r_0 = f - Ax_n^{\text{CI}}$, cf. (7.1.1)

Chebyshev Polynomials and Elliptic Functions

$\|\cdot\|_\Omega$	Chebyshev norm w.r.t. Ω
$\mathcal{T}_n(t; \Omega)$	Chebyshev polynomial w.r.t. Ω, cf. Corollary 3.1.4
T_n	Chebyshev polynomial of the first kind, cf. (2.1.1)
U_n	Chebyshev polynomial of the second kind, cf. (3.2.4)
$\mathcal{P}_n(t; \Omega, \xi)$	optimal polynomial w.r.t. Ω, cf. Corollary 3.1.4
$g(z; \Omega^c, z_0)$	Green's function, real, cf. (3.1.7)
$\mathcal{G}(z; \Omega^c, z_0)$	Green's function, complex, cf. (3.1.7)
sn, cn, dn	Jacobian elliptic functions, cf. (3.3.7), (3.3.8)

Notation

k	modulus, cf. (3.3.7)
$K(k)$	complete elliptic integral of the first kind, cf. (3.3.9)
H, θ	Theta functions, cf. (3.3.11)
$\varphi(u;\rho)$	conformal mapping, cf. (3.3.17)
$G(u;\rho)$	analytic function, cf. (3.3.22)
$\kappa(\Omega)$	asymptotic convergence factor w.r.t. Ω, cf. (3.1.5)

Stokes Problem

$L_2(\Omega), \langle \cdot, \cdot \rangle_{L_2}$	space of square integrable functions, cf. (8.1.1), (8.1.2)
$H^1(\Omega), \langle \cdot, \cdot \rangle_{H^1}$	Sobolev space, cf. (8.1.3), (8.1.4)
$X, \|\cdot\|_X$	velocity test space, cf. (8.1.9)
X_0	"incompressible" X, cf. (8.1.14)
Q	pressure test space, cf. (8.1.10)
X^h, Q^h	finite dimensional subspaces of X and Q, respectively
N_j, M_j	basis functions for X^h and Q^h, respectively
$a(\cdot,\cdot), b(\cdot,\cdot)$	bilinear forms, cf. (8.1.12)
$J(\cdot,\cdot)$	saddle point function, cf. (8.1.17)
$\mathcal{Q}_p$	pressure mass matrix, cf. (8.2.13)
$\mathcal{T}_h$	triangulation, cf. Section 8.3
$\mathcal{P}_n$	polynomial function space, cf. (8.3.1)

12 Index

Abel, 111

Achieser, 43, 93, 96, 105, 106, 116, 119, 122–124

Ahlfors, 93, 94, 100

Approximation of the A-norm
  CG case, 250
  general case, 259
  implementation of, 254

Ashby, 155

Asymptotic convergence factor
  definition of, 91
  for two intervals, 125
  implementation, two intervals, 130

Atlestam, 104

Axelsson, 235

Babuška, 230

Barker, 235

Bercovier, 243

Braess, 225, 231, 238

Brezzi, 225, 230, 232, 235, 242

Brodlie, 149

Butland, 149

Byrd, 105

Calvetti, 71

Carlson, 126, 149

CG method
  applied to indefinite systems, 166
  approximation of the A-norm, 250
  implementation of, 161
  operation count for, 161
  orthogonal polynomials and , 54

Chandra, 8, 60, 176, 178

Chebyshev, 99

Chebyshev iteration
  leapfrog variant of, 215
  optimal polynomial and, 212

Chebyshev polynomial
  of the first kind, 19
  Chebyshev iteration and, 212
  definition of, 91
  for one interval, 97
  for two intervals, 102

Cheney, 40

Chihara, 20, 56, 65

Christoffel number, 31

Compatibility condition, 230

Continued fractions, 37

Courant, 119

CR method
  applied to indefinite systems, 166
  implementation of, 164

Index

kernel polynomials and, 57
operation count for, 164
Craig, 158
Craig's method
  implementation of, 199
  operation count for, 199
Cullum, 88

Dahlquist, 261, 262
Davis, 55
de Boor, 125, 149
Deuflhard, 190, 251
Distribution function
  approximation of, 145
  definition of, 22
Donavan, 36

Eiermann, 8, 91, 92, 212
Eisenstat, 176, 261, 262
Elliptic function
  definition of, 110
  elliptic cosine, 112
  elliptic sine, 111
Elliptic integral of the first kind, 112
Elman, 246
Energy norm, 228, 249
Equilibrium distribution, 94

Finite element
  P_1 isoP_2 method, 242
  Q_1 isoQ_2 method, 243
  definition of, 237
  Hood-Taylor method, 242
Fischer, 32, 36, 71, 107, 125, 144, 210, 217, 246, 257, 260, 261

Fletcher, 79, 174, 176
Flop, 18
Fokkema, 171
Fortin, 225, 235, 242
Freund, 9, 63, 69, 79, 85, 125, 129, 142, 144, 174, 178, 180, 210, 217
Fridman, 70, 79, 174
Friedman, 105
Fritsch, 149

Galerkin condition, 54
Gaussian quadrature, 30
Gautschi, 26, 30, 36, 43, 65, 254, 260, 261
Geronimo, 106
Girault, 225, 226, 237, 242
Givens rotations, 45
Golub, 7, 18, 30–32, 36, 143, 161, 192, 210, 212, 237, 257, 258, 260–262
Golub/Kahan Bidiagonalization, 192
Grade, 16
Grcar, 96
Green's function
  definition of, 92
  for one interval, 94
  for two intervals, 119
Greenbaum, 88, 258
Gunzburger, 225, 237
Gutknecht, 155

Hackbusch, 225, 228, 230, 249
Hanke, 71
Harmonic Ritz values, 142
Helms, 94
Henrici, 93, 119

Hermite kernel polynomial
    Christoffel-Darboux identity of, 78
    definition of, 73
    extremal property of, 73
    ME approach and, 156
    minimal error method and, 75
    orthogonality of, 73
    recurrence relations of, 79
    zeros of, 83
Hestenes, 7, 54, 195, 203
Hilbert, 119
Hille, 94, 95
Hochbruck, 71
Householder, 155

Inconsistent system, 71
Inf-sup condition
    continuous, 230
    discrete, 232
Iteration polynomial, 15

Jacobi, 111
Jacobi matrix, 27
Johnson, 225
Joukowsky map, 93

Kahan, 192
Karlin, 36
Kent, 262
Kernel polynomial
    Christoffel-Darboux identity of, 65
    computation of zeros, 69
    CR method and, 57
    definition of, 55
    extremal property of, 56
    Hermite kernel polynomial and, 89
    minimal residual method and, 57
    MR approach and, 156
    orthogonality of, 55
    recurrence relations of, 60
    reproducing property of, 55
    zeros of, 65
Kober, 116
Krylov
    monic basis, 135
    orthonormal basis, 132
Krylov subspace
    definition of, 16
    method, 16

Lanczos, 8, 134
Lanczos method
    estimating the distribution f., 145
    estimating the spectrum, 143
    implementation, 134
Lebedev, 108
Li, 8, 92
LSQR method
    implementation of, 195
    operation count for, 195

Manteuffel, 69, 142, 155
Markoff, 98
MCR method
    implementation of, 176
    kernel polynomials and, 57
    modifications of, 178
    operation count for, 176
Meinardus, 91

Minimal residual smoothing, 189
Minimal solution, 42
Minimal true error approach, 189
MINRES method
  implementation of, 185
  kernel polynomials and, 57
  operation count for, 185
Model problem, 137
Modersitzki, 125, 204
Modified moments, 36
Modulus, 112

Nachtigal, 210
Nash, 262
Niethammer, 91, 212
Nodal basis function, 240
Node point, 239
Normal equations
  CGNE approach and, 199
  CGNR approach and, 195
  Craig's method applied to, 199
  LSQR method applied to, 195
  symmetric spectrum and, 207

Optimal polynomial
  Chebyshev iteration and, 212
  definition of, 90
  for one interval, 97
  for two intervals, 102
Orthogonal polynomial
  computation of zeros, 30
  extremal property of, 28
  monic version, 20
  of the second kind, 39

orthonormal version, 20
  three-term recurrence rel. of, 23
  zeros of, 29
Orthogonal residual polynomial
  CG method and, 54
  definition of, 17
  extremal property of, 51
  Hermite kernel polynomial and, 85
  kernel polynomial and, 87
  OR approach and, 156
  recurrence relations of, 48
Otto, 69, 142

Paige, 8, 60, 65, 79, 85, 142, 143, 157, 158, 167, 180, 195
Parameter dependent method
  based on eigenvalue distrib., 217
  based on eigenvalue estimates, 212
Parameter free method, 155
Parlett, 27, 65, 141–143, 167
Peherstorfer, 105–107, 109, 124
Perron, 21, 41
Petrov-Galerkin condition, 57
Peyret, 244
Pironneau, 243
Polynomial based iterative method, 15
Posse, 34
Pressure mass matrix, 235

QR factorization
  bidiagonal case, 194
  tridiagonal case, 44

Ramage, 210
Raviart, 225, 226, 237, 242

Rayleigh-Ritz approximation, 141
Reichel, 71
Reid, 60
Reinhardt, 144
Residual polynomial, 15
Rice, 125
Ritz values, 141
Rutishauser, 57

Saad, 180, 212
Sack, 36
Saddle point problem, 229
Sauer, 125
Saunders, 8, 60, 79, 85, 157, 180, 195
Saylor, 155
Schultz, 176, 180
Schönauer, 188
Shapley, 36
Silvester, 210, 236, 245, 246
Singular system, 71
Sleijpen, 171
Smirnov, 21
Spline, monotone
    definition of, 149
    implementation of, 150
Standard inner product, 18
Startresidual, 16
Stiefel, 7, 8, 54, 57, 164, 195, 203
Stieltjes, 37
Stieltjes procedure
    for kernel polynomials, 61
    for orthogonal residual p., 49
    for orthonormal polynomials, 26
    for polynomial weight function, 153
Stiffness matrix, 235
STOD method
    Hermite kernel polynomials and, 75
    implementation of, 174
    modifications of, 178
    operation count for, 174
Stoer, 9, 79, 174, 190
Stokes problem
    classical formulation of, 226
    discrete formulation of, 231
    variational formulation of , 227
Strakoš, 257, 258
Struble, 60
SYMMLQ method
    Hermite kernel polynomials and, 75
    implementation of, 180
    operation count for, 180
Szegö, 19, 29, 31, 37, 50, 65, 68, 153
Szyld, 85

Taylor, 244
Theta function, 114
Triangulation
    admissible, 237
    regular, 238

Uniform approximation, 96

van Assche, 106
van der Sluis, 141
van der Vorst, 65, 141–143, 167, 171
Van Loan, 18
van Loan, 143, 161, 237
Varga, 7, 8, 91, 92, 138, 212

Index

Walker, 188
Walsh, 92
Wathen, 210, 235, 236, 245, 246
Watson, 110, 112–114, 117, 129
Weight function, 19
Weiss, 188
Welsch, 30, 31
Wheeler, 36
Whitakker, 129
Whittaker, 110, 112–114, 117
Widlund, 85
Widom, 92
Wilf, 31
Wilkinson, 16
Wimp, 43

Zero-order moment, 26
Zhang, 71
Zhou, 188

MIX
Papier aus verantwortungsvollen Quellen
Paper from responsible sources
FSC® C105338

If you have any concerns about our products,
you can contact us on
ProductSafety@springernature.com

In case Publisher is established outside the EU,
the EU authorized representative is:
Springer Nature Customer Service Center GmbH
Europaplatz 3, 69115 Heidelberg, Germany

Printed by Libri Plureos GmbH
in Hamburg, Germany